# China's International Transboundary Rivers



China has 40 major transboundary watercourses with neighbouring countries, and has frequently been accused of harming its downstream neighbours through its domestic water management policies, such as the construction of dams for hydropower. This book provides an understanding of water security in Asia by investigating how shared water resources affect China's relationships with neighbouring countries in South-East, South, Central and North-East Asia.

Since China is an upstream state on most of its shared transboundary rivers, the country's international water policy is at the core of Asia's water security. These water disputes have had strong implications for China's interstate relations, and also influenced its international water policy alongside domestic concerns over water resource management.

This book investigates China's policy responses to domestic water crises and examines China's international water policy as well as its strategy in dealing with international cooperation. The authors describe the key elements of water diplomacy in Asia which demonstrate varying degrees of effectiveness of environmental agreements. It shows how China has established various institutional arrangements with neighbouring countries, primarily in the form of bilateral agreements over hydrological data exchange. Detailed case studies are included of the Mekong, Brahmaputra, Ili and Amur rivers.

**Lei Xie** is a Visiting Research Scholar, Institute of Geographic Sciences and Natural Resources Research, Chinese Academy of Sciences, Beijing, China. She receievd her PhD from Wageningen University, the Netherlands, after gaining her BA from School of Government, Peking University, China and MPhil from the Chinese University of Hong Kong.

**Shaofeng Jia** is Professor and Chair of the Department of Water Resources Research, Institute of Geographic Sciences and Natural Resources Research, Chinese Academy of Sciences, Beijing, China.

## Earthscan Studies in Water Resource Management

**Community Management of Rural Water Supply**
Case Studies of Success from India
*Paul Hutchings, Richard Franceys, Stef Smits and Snehalatha Mekala*

**Drip Irrigation for Agriculture**
Untold Stories of Efficiency, Innovation and Development
*Edited by Jean-Philippe Venot, Marcel Kuper and Margreet Zwarteveen*

**Water Policy, Imagination and Innovation**
Interdisciplinary Approaches
*Edited by Robyn Bartel, Louise Noble, Jacqueline Williams and Stephen Harris*

**Rivers and Society**
Landscapes, Governance and Livelihoods
*Edited by Malcolm Cooper, Abhik Chakraborty and Shamik Chakraborty*

**Transboundary Water Governance and International Actors in South Asia**
The Ganges–Brahmaputra–Meghna Basin
*Paula Hanasz*

**The Grand Ethiopian Renaissance Dam and the Nile Basin**
Implications for Transboundary Water Cooperation
*Edited by Zeray Yihdego, Alistair Rieu-Clarke and Ana Cascao*

**Freshwater Ecosystems in Protected Areas**
Conservation and Management
*Edited by Max C. Finlayson, Jamie Pittock and Angela Arthington*

**Participation for Effective Environmental Governance**
Evidence from European Water Framework Directive Implementation
*Edited by Elisa Kochskämper, Edward Challies, Nicolas W. Jager and Jens Newig*

**China's International Transboundary Rivers**
Politics, Security and Diplomacy of Shared Water Resources
*Lei Xie and Shaofeng Jia*

*For more information and to view forthcoming titles in this series, please visit the Routledge website: www.routledge.com/books/series/ECWRM/*

# China's International Transboundary Rivers

## Politics, Security and Diplomacy of Shared Water Resources

**Lei Xie and Shaofeng Jia**

Routledge
Taylor & Francis Group
LONDON AND NEW YORK

earthscan
from Routledge

First published 2018 by Routledge

2 Park Square, Milton Park, Abingdon, Oxfordshire OX14 4RN
52 Vanderbilt Avenue, New York, NY 10017

*Routledge is an imprint of the Taylor & Francis Group, an informa business*

First issued in paperback 2019



*British Library Cataloguing in Publication Data*
A catalogue record for this book is available from the British Library

*Library of Congress Cataloging in Publication Data*
Names: Xie, Lei, author. | Jia, Shaofeng, 1964– author.
Title: China's international transboundary rivers : politics, security and diplomacy of shared water resources / Lei Xie and Jai Shaofeng.
Description: Abingdon, Oxon ; New York, NY : Routledge, 2018. | Series: Earthscan studies in water resource management | Includes bibliographical references and index.
Identifiers: LCCN 2017030170 | ISBN 9781138689060 (hardback) | ISBN 9781315537900 (ebook)
Subjects: LCSH: Water use–Government policy–China. | Water-supply–Co-management–China. | Water-supply–China–International cooperation. | Water resources development–Political aspects–China. | Water rights–China. | Water security–China. | China–Foreign relations–Asia.
Classification: LCC HD1698.C5 X526 2018 | DDC 333.91/620951–dc23
LC record available at https://lccn.loc.gov/2017030170

ISBN: 978-1-138-68906-0 (hbk)
ISBN: 978-0-367-40370-6 (pbk)

Typeset in Goudy
by Wearset Ltd, Boldon, Tyne and Wear

To *Lili*

# Contents

# Preface

Water is a powerful element on the planet. It nourishes all life and shapes civilizations. Water features prominently in Chinese culture. Laozi, an ancient philosopher states in the Book of Tao and Theh, 'The top virtue is like water, benefiting all, contending with none, and is content with what all others disdain. Therefore it is close to the Word.'

China is one of the most water-rich countries in the world and benefits from 4,000 years' experience in the governance of water, with constant evolution in its approaches and philosophy. For a long time, management and control were a strong focus of China's approach to the use of water. This is also referred to as *Shuili*, which means drawing the benefits from water resources as well as avoiding the difficulties that water disasters could bring. Today, natural scientists and engineers play a significant role in water governance. Emphasis is placed on technical issues in water resource management and in the sharing of water resources across borders. However, as the contribution of social scientists remains doubted to this day, the central government has not yet established effective multilevel water governance that incorporates various actors.

This book is a cross-disciplinary study that has adopted subjects of politics, water resource management, international relations and area studies of China and Asia. By doing so, it aims to broaden our understanding of the effectiveness of collaborative mechanisms that are reached among riparian countries; it also provides insights on Asia's regional security, which is potentially threatened by ongoing water disputes between China and neighbouring states.

We are grateful for the assistance and support provided by Joshua Garland, Zha Daojiong, Yan Jiabao, Wen Bo and Yang Yong.

# Abbreviations

| | |
|---|---|
| ASEAN | Association of Southeast Asian Nations |
| BCI | Biodiversity Conservation Corridors Initiative |
| BDCA | Border Defence Cooperation Agreement (China and India) |
| BOT | built-operation and transfer |
| DBOA | Department of Boundary and Ocean Affairs |
| EIA | Environmental Impact Assessment |
| ELM | Expert Level Mechanism |
| EU | European Union |
| FDI | foreign direct investment |
| GBM | Ganges–Brahmaputra–Meghna |
| GDP | Gross domestic product |
| GEF | Global Environmental Facility |
| GMS | Greater Mekong Subregion |
| GMSM | Greater Mekong Subregion Mechanism |
| GW | gigawatt(s) |
| ha | hectare(s) |
| IWRM | integrated water resource management |
| kW | kilowatt(s) |
| kWh | kilowatt-hour(s) |
| LMCM | Lancang–Mekong Cooperation Mechanism |
| LMCs | Lower Mekong Countries: Vietnam, Cambodia, Lao People's Democratic Republic, Thailand and Myanmar |
| MEP | Ministry of Environmental Protection |
| MFA | Ministry of Foreign Affairs |
| MOU | memorandum(a) of understanding |
| MRC | Mekong River Commission |
| MW | megawatt(s) |
| MWR | Ministry of Water Resources |
| NDRC | National Development and Reform Commission |
| NGO | non-governmental organization |
| NTS | Non-Traditional Security |
| ODA | official development assistance |
| RFE | Russian Far East territory |

| | |
|---|---|
| SCO | Shanghai Cooperation Organisation |
| TEIA | transboundary environmental impact assessment |
| TRL | Three Red Lines |
| UN | United Nations |
| UNDP | United Nations Development Programme |
| UNECE | United Nations Economic Commission for Europe |
| UNEP | United Nations Environment Programme |
| UNWC | United Nations Watercourses Convention |
| US | United States |
| USA | United States of America |
| USSR | Union of Soviet Socialist Republics |
| WMI | water management institutions |
| WWF | World Wide Fund for Nature |

# 1 China's transboundary rivers: politics, diplomacy and security

## An introduction

### Introduction: an understanding of water cooperation

Conventional wisdom suspects that competition over water is a source of conflict at varying levels of intensity. According to Homer-Dixon (1999), ecological impacts have strong adverse effects on human populations and result in competition for resources which can lead to national conflicts as well as violence between countries. Yet, methodologically, that body of work on environmentally induced conflict has so far only produced findings that lack the support of systematic empirical evidence (Gleick, 1990; Homer-Dixon, 1994; Butts, 1997). We are thus informed that countries are more likely to react to water disputes with cooperation than to fight for control and management of natural resources (Wolf, 1999; Brochmann and Hensel, 2009). Theories of international relations including realism and neo-liberalism have been adopted as theoretical tools to understand why countries are incentivized to form water agreements. However, both of these theoretical concepts have limitations in explaining the complex conditions upon which formal agreements are reached (Tir and Ackerman, 2009).

Indeed, the understanding of water cooperation has been contentious. In studies of water conflict, *water cooperation* refers to the negotiating process that countries engage in to soften disagreement and resolve conflicts over the use of shared water resources. It is recognized that diplomatic initiatives play a significant role in mitigating interstate disputes over water resources. A visible simplified dichotomy is often found of the term *water cooperation* (Zeitoun and Mirumachi, 2008). For example, conflicts are equalized as events of political tensions, or extreme conflictive events (Tir and Ackerman, 2009), and according to Gerlak and Grant (2009), cooperation is viewed as an opposite position to conflict. Moreover, some authors argue that cooperation and conflicts are hard to distinguish, as they can occur simultaneously in negotiating over how river basins are shared (Wolf *et al.*, 2003; Zeitoun and Mirumachi, 2008; Zawahri and Gerlak, 2009). Certain issues around water management are identified as sources of tension. And the major causes of tension constitute, for instance, competition over freshwater supplies (Wolf *et al.*, 2005), navigation (Brochmann and Gleditsch, 2006; Collier, 2007) and the deterioration of water quality in shared river systems (Yoffe *et al.*, 2003).

It is noteworthy that lacking a cross-disciplinary approach, single-disciplinary research has not yet succeeded in providing solutions to solving issues related to international rivers that run through two or more countries. Studies of international politics focus on the nature of water conflicts as a potential trigger to exacerbate friction in interstate relations. Such a view has been criticized for being too 'linear and incremental' (Biswas, 2011: 663), and is now being challenged in light of the nature of water conflicts of varying scales of intensity (Bernauer and Kalbhenn, 2010; Biswas, 2011). Natural scientists and engineers primarily focus on watershed analysis, international law principles or technical issues in cooperation (Tolba and Rummel-Bulska, 1998; Biswas, 2011). Based on rational choice perspective, they assume that individual states are merely incentivized by scientific judgement, neglecting situations where political interests are combined in states cooperative initiatives. While they contribute greatly to the sustainable management of domestic water resources, they are impeded by incorporating little consideration of how foreign relations impact on state behaviour and their preferences in water management.

Water has a multifaceted nature and is perceived in various ways. Water cooperation is a complicated arena where domestic politics and international relations intersect, it involves a complexity of features, and it may be vested with various interests. Therefore, an open mind is required to understand the complex features of water cooperation.

## Water cooperation in Asia

Transboundary water courses, also termed international rivers, are rivers that run across territorial borders. They abound in Asia and are shared among many countries, in Afghanistan, Bangladesh, Bhutan, Cambodia, India, Kazakhstan, Kyrgyzstan, Laos, Mongolia, Myanmar, Nepal, North Korea, Pakistan, Russia, Tajikistan and Vietnam. River basins, such as the Mekong basin, represent resource-rich hotspots for riparian states. These states are also facing growing pressures to react to global change, particularly in the forms of industrialization, climate change and increasing demands for ecosystem protection (Varis *et al.*, 2012). Indeed, increasing pressures are being exerted on water resources from a high level of both poverty and population growth, and from rapid urbanization. Most countries are experiencing difficulties in the sustainable management and development of water resources, which are used in cross-sectoral ways for environmental protection, economic growth, and the production of food and energy. Vulnerability in the river basin is unevenly spread among riparian states (Gain and Schwab, 2012).

Issues of managing international rivers in Asia have been heatedly debated. Concerns have been raised over the approaches that Asian countries are taking to develop and sustainably manage water basins. Most of all, competition over freshwater supplies is thought by many to be a cause of violent conflicts, and this notion has inspired considerable scholarship. Although very weak causal links have been found between water-related armed factions and interstate hostilities (Bernauer and Kalbhenn, 2010), contextual factors in Asia have raised the questions of if, and to what extent, countries are likely to react to water disputes through cooperation rather than conflict for the control and management of natural resources. Various conditions are recognized as impacting on the potential for countries to cooperate in the sharing of water basins. These contextual factors have raised the questions of if, and to what extent, countries can succeed in reacting to water disputes through diplomatic cooperation rather than conflict for the control and management of natural resources. Some of the major issues identified are outlined next.

Regional security in South Asia and South-East Asia has been found to be unstable and rapidly changing. In this region, power relations are generally unstable, with historical mistrust and various political frictions, such as border disputes. Water disputes are likely to worsen political tensions among adjoining riparian states that share water resources. International politics have impacted on the mapping of major rivers. For example, in North and Central Asia the creation of new states following the disintegration of the former Soviet Union has increased the number of international watercourses (Biswas, 2008). The newly independent states are left in the void of effective institutional arrangements and are faced with increasing threats relating to conflicting interests and political tensions.[1]

Developing countries are noted to be lacking the commitment to international environmental norms which facilitate interstate cooperation (Bernauer and Kalbhenn, 2010). Agreeable foreign relations are a prerequisite for regime formation (Haftendorn, 2000; Brochmann and Gleditsch, 2012). However, most of the riparian states are also not party to major global water norms, but remain keen to cope with water conflicts through bilateral foreign relations.[2] Bilateral agreements over transboundary river basins have been the primary form of cooperation among neighbouring riparian states. For instance, no clear programme to develop collective water management practices has been initiated among the Brahmaputra riparian countries, the Ili River basin, nor the Amur River basin. The Ganges–Brahmaputra–Meghna (GBM) River system is regarded as one of the least coordinated in the world (ICA, 2012). It has been suggested that authoritarian states are less likely to join cooperative mechanisms to address water disputes and that this is likely to lead to worsened political tensions with their adjoining riparian states (Brochmann and Hensel, 2009; Gerlak and Grant, 2009). This may also apply to other Asian countries with restrictive regimes. It is therefore not surprising that the governance of Asia's transboundary river basins is rather weak (Varis *et al.*, 2012).

The weak capacity of many neighbouring riparian states is another important factor affecting Asian countries' formation of cooperative efforts. These countries are likely to be facing concerns around water management, including the pressing needs to improve water management capacity and development outcomes, as well as requirements to improve economic, institutional and technological instruments (Biswas, 2011). Additionally, the above-mentioned riparian countries are faced with challenges in effectively establishing multiple levels of governance arrangements that are crucial to realizing sustainable development (Dietz *et al.*, 2003; Raadgever *et al.*, 2008). With their capacity to govern water and to respond to increasing environmental change being questioned, these countries are hypothesized as being likely to be involved in interstate conflicts over natural resources (Homer-Dixon, 1994).

## Water disputes and China's diplomacy

China has rich water resources, which includes 42 major transboundary watercourses (including lakes) that are shared with neighbouring countries in the north-west, north-east and south-west of the country. These countries are Afghanistan, Bangladesh, Bhutan, Cambodia, India, Kazakhstan, Kyrgyzstan, Laos, Mongolia, Myanmar, Nepal, North Korea, Pakistan, Russia, Tajikistan and Vietnam. This ranks China after Russia and Argentina as the third nation (together with Chile) with the largest number of international rivers (Liu *et al.*, 2006). Among China's rivers, 15 are among the world's longest rivers. Twelve of these rivers originate in China, and in some cases there are multiple watercourses shared with countries that include Russia, Mongolia and Myanmar. China is upstream of most shared transboundary rivers. Among these, water disputes have been identified about water quality or quantity, namely the Yalu, Tumen, Amur, Suifen, Irtysh, Ili, Tarim, Indus, Ganga, Brahmaputra, Irrawaddy, Salween, Mekong, Red and the Pearl Rivers, which are regarded by the Chinese as the most important (see Figure 1.1) (He *et al.*, 2014). More specifically, international rivers

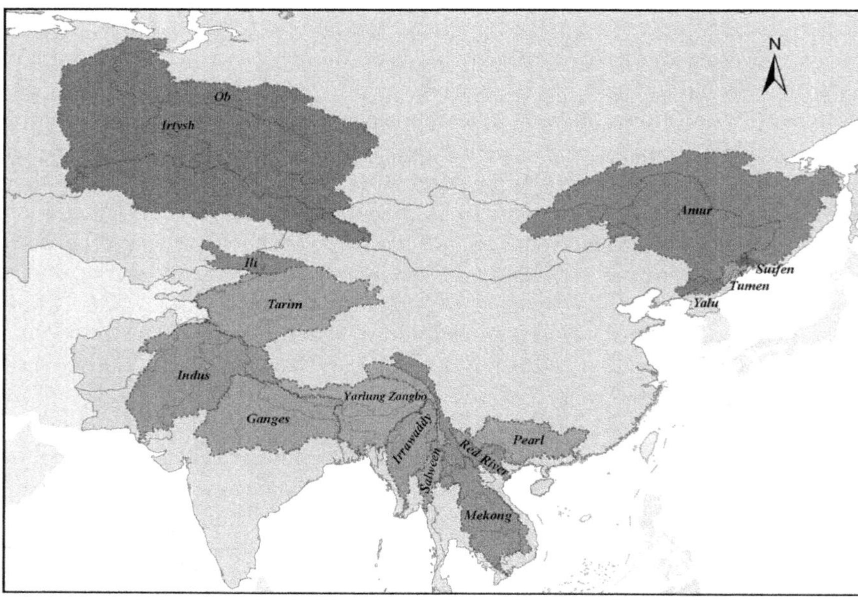


*Figure 1.1* China's transboundary rivers.

Source: authors' commission, credit: Jiabao Yan.

in the north-east are characterized as border rivers;[3] in the south-east, transboundary rivers originate from both within China and from outside the country; while in the south-west, most originate from within the country.

### *Foreign policy and diplomacy*

The issue of sharing water has raised disputes and sometimes tensions between China and its lower riparian neighbours, some of whose inhabitants are among the poorest in the world (Khan, 1996; Bandyopadhyay and Ghosh, 2009). Being located upstream, China, on various occasions, has been accused of taking unilateral actions that cause damage to downstream countries' interests such as constructing dams or water diversion projects (Chowdhury, 2010; Sood and Mathukumalli, 2011). These water disputes are regarded as a source of conflict between China and its downstream neighbours. Given their status as developing nations, these neighbouring countries may experience particular challenges to resolving this conflict. With China actively seeking to protect its own interests by accessing and exploiting water resources, its water interactions with neighbouring countries are regarded as having the potential to spark conflict at the interstate level.

Nevertheless, varying degrees of diplomatic initiatives have been formed between China and its neighbouring riparian states. This is primarily confirmed by developing water treaties in the form of agreements or memoranda of understanding (MOUs) (see Table 1.1). Joint bodies have been established to implement agreed resolutions. This is seen in water agreements with Mongolia, Russia, Kazakhstan and India. Lately, China has proactively promoted the formation of the Lancang–Mekong Cooperation Mechanism (LMCM), which is based on the multilateral mechanism of the ASEAN organization and serves to

*Table 1.1* China's bilateral mechanisms on water

| *Country of interaction* | *Number of agreements* | *Water information* |
|---|---|---|
| Russia | 13 treaties | Hydrological information |
| North Korea | 11 treaties | Ibid. |
| Kazakhstan | 9 treaties | Ibid. |
| Mongolia | 4 treaties | Ibid. |
| Myanmar | 4 treaties | Ibid. |
| Laos | 3 treaties | Ibid. |
| Vietnam | 2 treaties | Ibid. |
| India | 3 MOUs | Ibid. |
| Afghanistan | None | Ibid. |
| Bhutan | None | Ibid. |
| Kyrgyzstan | 2 treaties | Ibid. |
| Nepal | None | – |
| Pakistan | None | – |

Note
MOUs = memoranda of understanding.

coordinate water sharing over the Mekong. This mechanism is designed by China with the support from the Asian Infrastructure Investment Bank. For the first time, an institutional cooperation has formed among all of the Lancang–Mekong River riparian states. These states include China, Myanmar, Laos, Thailand, Cambodia and Vietnam. With China's involvement, the multilateral mechanism has, for the first time, resulted in state-led cooperation over the Lancang–Mekong River.

According to Chen *et al.* (2013), these institutional arrangements are found to support and follow some of the key principles of global water norms such as the exchange of data and information between relevant states and with the Mekong River Commission. In most of the cases in which institutional cooperation has been formed, information provision and data exchange is identified. However, the Chinese government has been cautious in the publication of information on its water governance. In particular, a clear concern in China's international water policy is peacemaking and the prevention of ecological disasters such as flooding and drought, or water quality degradation. This is referred to as a form of emergency response cooperation. For example, China has formed such cooperation with Mongolia, Kazakhstan and Russia.[4] With India, China has developed progressive diplomatic initiatives on the issues of 'natural disasters'[5] and 'emergency management'[6] when disputes have arisen, stating that China's unilateral dam construction upstream was the source of a disaster, Chinese officials attempted to ease the tensions by providing dry-season water level data. This data was also aimed at helping with disaster preparedness including the flash flood guidance system and flood forecasts in the Mekong River for improved navigation, drought forecasting and water resources management. In such institutional cooperation, China has agreed to share more hydrological data with the Mekong River Commission by extending the hydrological data provision by 30 days, starting on 1 June until 31 October, every year, as well as increasing the frequency of the data sharing to twice a day. Nevertheless, complaints on the scope of data shared between China and its riparian neighbours have been received (although the country has a low record of complaints). Overall, however, China's compliance with the water agreements has been confirmed, with all of the international water agreements it is involved in (Chen *et al.*, 2013).

### *Domestic water policy and governance*

Water has a multifaceted nature and China's perception of water is multifaceted. Water is regarded as one of the nation's key natural resource assets, of great significance to China's development. Overall, China is moderately secure in terms of its water resources (Figure 1.2). Severe water stress, in the form of annual average stress (1981–2000), has been found only in Northern China (Varis *et al.*, 2012; Jia *et al.*, 2014; Kattelus *et al.*, 2015). Facing increasing demands for energy and environmental services, China's transboundary water resources is believed to contribute to meeting this growing demand.

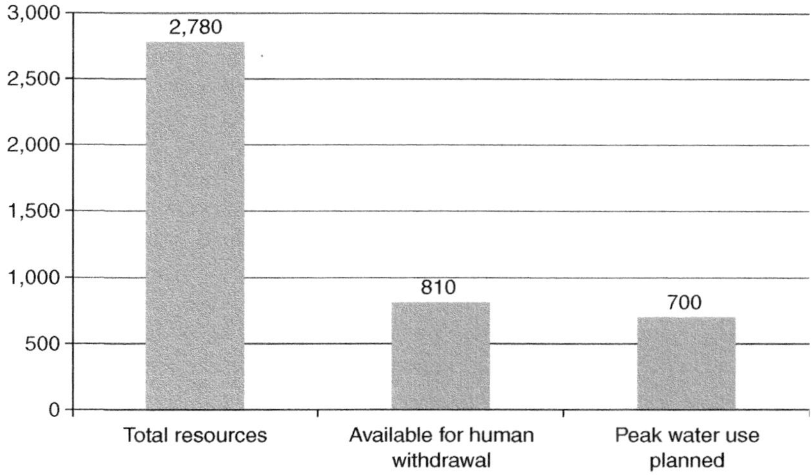


*Figure 1.2* China's water resources and water use.
Source: Jia *et al.* (2014).

China has displayed a clear technocentric perspective in water management. Rooted in its national identity, it is argued that the Chinese government has shown its decisive attitude towards the non-human world and has given little credence to environmental security (Shapiro, 2012). It is firmly believed that humans are capable of reacting to environmental changes, hence, the perception of an environmental threat is assessed by the capacity of a country's environmental management. In water management, this is illustrated with an aim to appropriate the resource by balancing the maximum level of economic profits with the minimum level of protection of aquatic ecosystems.

Along with China's economic growth and increasing urbanization, its evolution in water management policy has been dynamic. Since 1949, the country has developed selected areas in water management. Not unlike those found in other developing countries, China's economic interests in water primarily concern irrigation, the exploration of hydroelectric power and other areas where economic profits can be generated. Only since the early 1990s has China developed a holistic water management system that incorporates ecological protection (Jia, 2013; Shen, 2014). In 2012, the central government promulgated the Three Red Lines policy, focusing on water resource protection, the improvement of water quality and an increase in water-use efficiency.[7] This unified regulation system of water resource management established quantitative goals for the efficiency and intensity of water use for the decade of 2020–2030. According to this policy document, China's national total water use is to be controlled within 700 billion $m^3$; its water-use productivity should reach or be close to average water use globally; and its water-use intensity of industrial added value should be reduced to less than 40 $m^3$. Hence, this is referred to as an 'important milestone' in China's strategic water development (Liu *et al.*, 2013). This policy document also has implications for the management of international rivers. It is inferred that this policy signals that domestic water resource management, including demand trends and the degradation of ecosystems along the international rivers, are intertwined and should be more broadly considered in China's policies on the management of international rivers (Feng *et al.*, 2015).

China's water governance features a centralized institutional arrangement that sees government authorities playing a dominant role in decision-making and policy implementation. Although multilateral governance arrangements are being developed, such practices often represent policy experimentation in certain contexts, mostly in wealthier regions (Silveira and Richards, 2013). Nevertheless, due to issues of unequal regional development, central authorities are increasingly challenged to govern water more effectively at the national scale. Indeed, water has become a source of instability across the nation. At the central level, policy negotiations over the sharing of water resources and the distribution of water quantity have been increasing (Moore, 2014). Concerns about pollution, apart from air pollution, constitute another key environmental welfare issue for the Chinese people (van Rooij, 2006; Xie, 2009).

Domestically, environmental governance is seen as being strongly monopolized by the central state, thereby limiting the sharing of environmental information between the government and non-state actors (Sonnenfeld and Mol, 2002). The Chinese government dominates environmental governance and environmental protection organizations are located within the state's administrative structure. They are not independent and do not possess authority over other functional units at different administrative levels. This framework suits coordination within a single industrial sector, but works poorly when cross-sectoral interests are combined (Ma and Ortolano, 2000; Zheng, 2001), especially when the Chinese Communist Party is involved (Jahiel, 1998). With the current Chinese environmental governance structure, the legitimacy of environmental policies has become increasingly challenged. Hence, in managing water resources in China, challenges lie not only in institutions and policy instruments, but also in the lack of effective implementation of water resource management that is based on intensive involvement of the public (UNEP, 2012). In particular, in the management of cross-sector water resources, increased pressure is also applied by the changing state–society relationship that affects the governance of various river basins, further threatening the effective and equitable management of transboundary river basins (Kattelus *et al.*, 2015). However, in the discussion of China's international water policy, little attention is placed on the link between foreign policy and its domestic roots. The debate has largely neglected China's domestic interests and environmental management, which are generally overshadowed by the implications that the countries' foreign policies have for transboundary issues (Mushkat, 2010).

### *Water as a strategic concern*

China's multifaceted perception of water is also reflected in the country's shifting attitude towards forging cooperation over transboundary water management (Hill, 2015). Cross-border water management represents a policy arena where domestic politics and international relations intersect. Closely linked to the internal and external faces of China's water policy, two dimensions are most discernible – consideration of domestic water management on the one hand; and on the other that water is incorporated into the development of the country's international security agenda. China, being aware of its geographical location and how the nation is perceived internationally, has made gradual changes in its foreign policy agenda towards dealing with water disputes. In particular, increasing attention has been given to transnational policy issues that show spillover effects beyond the country's national borders (Feng *et al.*, 2015).

More specifically, water as a natural resource has become a security interest in China's foreign policy agenda. The notion was first developed around 2002 under Jintao Hu's administration, and further elaborated in 2013 under Jinping Xi's administration. In comparison to Hu's conceptualization, Xi's notion of non-traditional security (NTS) is deepened, reflecting the changes China is making to consider human security in various contexts. According to Xi, NTS refers to energy resource security and major natural disasters along with other concerns such as 'terrorism, transnational crimes, environmental security and cyber security'.[8]

China's strengthening of the concept in its foreign policy is interpreted as the country's intention to maintaining trustworthy with neighbouring riparian states. In such process, the country has become more willing to commit to peacemaking, and actively respond to water-related disasters that pose challenges to human security (Breslin, 2015), including drought, flooding and those that affect public health. By politicizing the water issue, commitments to international peacemaking may represent an escalated security act that helps to promote a stable regional security environment.

Nevertheless, China's perception of water issues is likely to be dynamic. By analysing China's foreign policy agenda, Zeng *et al.* (2015) suggest that the significance of international environmental policy may change. It is foreseeable that transboundary environmental issues, represented by transboundary water sharing, may constitute one of the key topics in the debate of China's core interests (Zeng *et al.*, 2015). However, a dearth of scholarship exists to provide insights into the conditions when such situations arise.

## Water security and China's rationale: research questions

Water is a key area that reflects a mixture of both instrumental interest and political intentions. China has shown an increasing interest in diplomatic initiatives with concerns expressed about whether the country and its riparian neighbours can overcome water-related tensions without resorting to open conflict. Currently, this is being achieved by China signing various forms of institutional arrangements with its riparian neighbours. However, as previously mentioned, the country has various vested interests in sharing waters with its riparian neighbours. Questions are raised on the principles, outcomes and rationales adopted in such diplomatic initiatives. And most of all, in a region like Asia, how China's water diplomacy will impact on already unstable foreign relations remains unanswered. This section elaborates on how these questions are developed.

## A *To what extent is China likely to adopt global water norms?*

### *International water norms and the greening of state sovereignty*

In both theory and practice, legal studies have started the discussion on normative power that the existing global norms have over interstate cooperation. Currently, international norms have been developed as a codification of state practices in response to hydrological interdependence, and how it has been interpreted and implemented with respect to shared water resources. On a global scale, there has been heated debate among policymakers and legal practitioners about how to govern shared river basins. Two legal frameworks have been influential in guiding countries to use transboundary waters and ensure their sustainable management, while preventing, controlling and reducing negative impacts on other nations. The Convention on the Protection and Use of Transboundary Watercourses and International Lakes was adopted in Helsinki in 1992, adopted by the Economic Commission for Europe (UNECE) and entered into force in 1996.[9] It is known as the Helsinki Convention, or the Water Convention. This Convention was originally a legal framework for transboundary water cooperation in European countries. Since 2013, countries outside the EU have been allowed to adopt it, making available a framework for wider use, globally.

On 21 May 1997, the Convention on the Law of the Non-Navigational Uses of International Watercourses (UN Watercourses Convention or UNWC) was adopted. It is one of the most influential laws on the global governance of transboundary water resources. However, the current practice of the UNWC has not yet been widely recognized as a framework providing basic provisions on water issues (Salman, 2007). Despite being opened in 1997 following a vote by 103 countries, few have since signed this convention.[10] It was not until 2014 that 35 countries ratified this convention and it was endorsed by the General Assembly of the United Nations and officially entered into force as a global legal document on transboundary water management.[11] On the whole, with the Water Convention, the two norms are complementary but the UNWC provides more detail than the former, particularly on the implementation of the law.[12]

As a rich body of soft law, the UNWC incorporates a range of instruments from internal and external dimensions. Core to International Water Law is the general obligation for cooperation between neighbouring riparian states, and potentially through the development of joint water institutions in the form of multilateral and cross-sectoral arrangements. Addressed in UN Watercourses Convention Articles 11–19, this principle contains procedures for countries to fulfil duties of prior information sharing and consultation. The broader goal lies in achieving equitable use and mutual benefits from the shared water resources. According to UNWC Article 20:9, 'Watercourse States shall, individually and, where appropriate, jointly, protect and preserve the ecosystems of international watercourses.'[13]

It should be noted that primary instruments have been adopted at international, European and basin levels (Moynihan and Magsig, 2014). Investigating the formation of water treaties in South Asia, legal scholars have concluded that a legal framework requires more contextualization as water governance is 'highly contextual and excessively dynamic in nature' (Uprety and Salman, 2011: 656). These authors concluded that findings of water cooperation drawn from liberal democratic political contexts may not always apply to different conditions (Mason, 2008). Hence, importing water laws that have proven successful elsewhere may not be a success for solving water disputes in Asia and Africa (Uprety and Salman, 2011).

### *China's attitude towards global norms*

Being a leading player in the development of clean technologies, China has shown substantial capacity and knowledge in developing hydropower. When China indicates a willingness to invest, national governments in industrializing countries can find it hard to resist the potential profits to be gained by jointly exploiting hydropower generation (Liebman, 2005; Zha, 2015). Studies have recognized a combination of incentives for China to engage in diplomatic cooperation and there is agreement that China's position towards UNWC and the UNECE remains ambiguous.

China has shown reservations in adopting global water norms, among which the UNWC is the most influential body of law on the governance of transboundary water resources. These laws advocate the formation of international cooperation among riparian states. For instance, the UNECE advocates the obligation to not cause harm (Salman, 2007: 8–9) and equitable and reasonable utilization of water. Concerns lie in China's joining of global water norms, which raises questions on China's exercising of its sovereignty. Experts, academics and policymakers have shown increasing concerns about this policy issue, which is contentious. To many policy elites, the country is poorly protected in exercising its sovereignty rights (He *et al.*, 2014) in relation to its utilization of water resources as an upper-stream state.

China has shown its contextualized understanding of the above norms. China interprets 'no harm' by assuming that harm will not occur if it incorporates scientific development strategies in its appropriation of transboundary water resources. China's Ministry of Foreign Affairs stipulates that 'new projects in transboundary rivers management must go through scientific planning and study, with the consideration of the interests of both downstream and upstream riparian countries' (He *et al.*, 2014). According to various authors (e.g. Caponera, 1987; McIntyre, 2014; Tarlock, 2014), in China's appropriation of water resources, the country commits to this norm and displays responsibility in reducing environmental harm by adopting environmental impact assessments in the government's planning and management of water.

Nevertheless, as previously mentioned, it remains unclear whether the existing water norms provide instrumental guidance to developing countries' sharing of rivers. To what extent the country is adopting global norms in appropriating transboundary water resources also remains unclear.

### B *What are the outcomes of China's diplomatic efforts to resolve conflicts and promote sustainable water management?*

China and its neighbouring riparian states have experienced increasing tensions. Scholarship of water disputes informs us that water resource management results from long-term and durable cooperation (Schmeier *et al.*, 2016) that can be reached through a broader scope of political dialogue (Carius, 2006: 14–15, 20; Beck, 2015: 220). In the case of forming cooperative partnerships for water resource management, riparian actors' interests in water also represent their mixed intentions, depending on the extent to which they rely on the shared water resources. Vested interests are recognized in developing diplomatic information-sharing collaboration with China. Collaborating countries responses to China's water diplomacy reveal dynamic changes in their bilateral policy that in turn reflect China's domestic policy agenda. Their strategic interactions over water may be different and (sometimes) conflicting. Hence, it becomes highly tangible if upstream countries' diplomatic cooperation results in the building of trust and 'harmony' with the other nations that it intends to cooperate with.

In the case of China, one potential issue is that expectations exist for China as a superpower with associated obligations. When the country's status quo as a superpower is stressed, we are reminded that we are presented with a paradox. In particular, superpowers need to appear responsible and rule-following in their pursuit of self-interest, since positive recognition by the international community is suggested as a key marker of a good state (Wolf, 2011). The more powerful the state, the greater the expectations of that state's virtuousness (Gilley, 2013). To many international observers, the future of Asia's regional security is uncertain, with questions around possible water disputes being a particular pressure point and catalyst for regional insecurity. With the varying nature of water diplomacy displayed by China, uncertainty exists whether China's strategic cooperation may lead to discord and disputes. In the context of Asia, where regional security is unstable, to what extent conflict resolution and water-basin management can be incorporated in water cooperation remains to be answered.

## C *What is China's rationale in forming water cooperation?*

China has shown ambiguous attitudes in response to water disputes in Asia. An upstream country, China does not always act rationally. However, it puzzles many why China has established diplomatic institutions with its neighbouring countries. Possessing geographical advantages implies that it is able to maximize its self-interest by taking unilateral actions. Three major international relations theories provide insights to understand China's rationale in forming such interactions. Each perspective describes certain features in how and why the country has taken such actions, it is hard to explain adequately the ambiguity that the country has shown in its water policy.

From a realism perspective, China's strategic acts do not serve instrumental rationality to advance national interests. It is stressed that, being upstream on most of its shared transboundary rivers, China possesses an advantage in managing shared water resources to enrich its material capacity as a superpower. Such a view focuses on China as being committed to gaining economic profit by exploiting shared water such as through dam construction and the development of hydroelectricity (Liebman, 2005; Ho, 2013; Zha, 2015). Further, Chellaney (2011) assumes that the geographical location China possesses allows it to act as hegemon towards downstream states to gain direct control over lower-stream nations. China's primary concern is economic profit. For instance, in some cases such as the Mekong River basin, it is recognized that Beijing exhibits an intention to enhance opportunities to develop the country's West and its subregions (i.e. Yunnan and Guangxi) (Hensengerth, 2009; Magee, 2012; Biba, 2013). Hence, domestic water resource management has been an important factor in motivating the country to instigate diplomatic cooperation with neighbouring countries.

From a liberalist perspective, China's domestic management needs have been overshadowed by the implications the country's policies have for transboundary issues. The nation's domestic interests have received insufficient attention (Mushkat, 2010: 516). Domestically the country has shown weakness in the securitization of freshwater resources (Biba, 2013). Although China occupies an upstream location on most of its transboundary rivers, the country has shown an intention to maximize its interests in the management of water basins. China may be seen as a leading actor that is capable of providing technical assistance and materials. Dam construction is a clear case in point (Liebman, 2005).

A constructivism perspective looks into the dissemination of international water norms into China, as impacted by international actors. Scheumann and Hensengerth (2014) examined standards and practices (on the production of sustainability standards) in water norms that are advocated by the World Commission on Dams. Taking the case of China, it is argued that although the country, along with India and Turkey, has rejected incorporating the World Commission on Dams and its decisions on sustainable development and decision-making in relation to dams, it has a strong incentive to engage and shape international norms. Following this line of analysis, China has shown concerns in responding to water-related emergencies. Studies of Chinese international relations have identified that China assumes a 'responsibility' that is perceived locally. This indicates that, as a superpower, it is obligated to implement its regional responsibility for security and wealth (Xiao, 2007). And such foreign policy interests have been identified in China's environmental peacemaking events, preventing natural disasters and mitigating their negative effects (Yeophantong, 2013). Unfortunately, there is a dearth of scholarship on how China interacts with other nations in its peacemaking activities over water.

Therefore, mixed interests are identified in China's water policies, which are recognized as short-term policies that suit the growing demand and changing interests the country exhibits in development (Biba, 2013). As Lowi (1993) notes, for a hegemon country like China (which possesses geographical advantage), it is equipped to appropriate transboundary water resources without forming institutional cooperation. It is inferred that China's strategic interests may not lie solely in the exploitation of water issues, but also with the potential tensions that it may be involved in.

Additionally, regional security among certain Asian countries has been found to be unstable and rapidly changing. China, being a superpower in Asia, plays a significant role in regional security and prosperity. Although China is perceived as a threat, discord between China and its riparian neighbour states is not immediately obvious. Moreover, triggered by tensions over the South China Sea, the sensitive geopolitical situation in Asia can hardly sustain any more frictions over foreign relations. Against this background, it is likely that China has developed strategic diplomacy and a complex agenda in its appropriation of transboundary waters (see Li, 2015). However, no systematic investigation has been carried out to understand the country's rationale and commitment in the complex transboundary rivers policy arena. It is essential that China's intentions and strategic interests are examined in order to understand the country's behaviour over water disputes.

## Method

This book is intended to provide an understanding of water security in Asia by investigating how shared water resources affect China's relationships with neighbouring countries. It aims to investigate China's policy responses to domestic water crises; examine China's international water policy as well as its position in coping with international cooperation; explore the nature of water cooperation in Asia, which sees varying degrees of effectiveness of environmental agreements; and contribute to the understanding of how effective collaborative mechanisms among neighbouring riparian countries can be reached. Although China is entitled to exercise its permanent sovereignty rights over rivers, its geographical location and regional power status has affected how it performs in this regard. The country's differing responses to this dilemma can have strong implications for the governance of transboundary waters as well as regional security.

There is a dearth of scholarship that provides cross-disciplinary work to investigate the complex intentions that underlie Chinese water diplomacy. The current research has been greatly influenced by studies on water disputes, either from the perspective of international relations or of water resource management. As a result of a narrowed perspective, the understanding of Chinese diplomacy is stagnant. Focus has been placed on institutional cooperation, without dynamic examination of the effects of such diplomacy. Water conflict and cooperation coexist as a continuum in interstate interactions (Goldstein, 1992; Warner and Mirumachi, 2008). In the Chinese case, the formation of water institutions does not necessarily signal that Chinese foreign policy is likely to be accepted by neighbouring riparian states. China's diplomacy constantly changes and the consideration of water is combined with considerations of the country's regional security. Hence, it would not be realistic to expect that substantial policy output over sustainable water management will result from diplomatic cooperation.

As witnessed in other regions, Asia's diplomatic water cooperation has displayed complex processes of formal cooperation (Tir and Ackerman, 2009). Nations' behaviours in diplomatic cooperation for a range of political reasons other than water management remain key drivers for the countries to form cooperative agreements. China has illustrated an evolving perception of water, which is weighed differently in the country's foreign policy agenda. Reflected in Chinese diplomacy, China's water interests are dynamic and in accordance with its desired outcomes. Hence, a process-based perspective should be adopted to understand Chinese diplomacy over shared waters. Similarly, for other nations, water interests are combined with other national interests in their cooperative behaviours with China. Depending on a country's reliance over shared water (Ovodenko, 2014), water may very well become a bargaining chip in these countries' negotiations with China.

## Structure of the book

Chapter 1 serves as a general introduction to the subject. Chapter 2 will outline how China's international water policy reflects its concerns about issues and reforms that have been undertaken in domestic water governance. China's strategy towards water management has evolved tremendously, as the country's transition to an industrialized nation created dramatic changes in its water usage. By analysing national water uses, it points out that the country has enjoyed a secure level of water use ratio on most of its rivers, including those that are shared with its riparian neighbours, and its overall water security has strongly influenced how the country appropriates water that it shares with its neighbouring states. The central authorities have begun to review its fundamental approach to accessing and exploiting water resources. This includes dramatic reforms that have touched upon the combined policy areas of water planning and resource management.

Chapter 3 examines how China exercises its sovereignty rights in the planning and management over international rivers. China is recognized as increasingly involved in the governance of the global environment, by committing to diplomatic cooperation on key environmental issues. Controversial views exist on principles and positions shown in China's international water policy. This chapter discusses the extent to which China adopts international water norms in its appropriating of transboundary water resources. It investigates the extent to which the country is committed to promoting water governance not only as result of rapid economic growth and industrialization, but also as an expression of assertiveness arising from the regional and global power that China is now able to assert over issues pertaining to the nation's sovereign rights.

In Chapter 4, a theoretical framework is developed which adopts a process-focused perspective to investigate China's complex water diplomacy with neighbouring countries, by examining its nature, forms and outcomes. In order to understand the complex nature of water cooperation, conceptual tools on the formation and features of water management institutions (WMIs) are analysed in order to better understand states' diplomatic initiatives when sharing transboundary rivers. It is highlighted that water represents a new arena where unique strategic interests and approaches are displayed in the policy of international water sharing. A case-study method is adopted. This deliberate methodology enables an investigation of contextual factors in such processes, including varying national interests in water resource management as well as the state of foreign relations exhibited between China and neighbouring states.

Chapters 5–8 present a range of case studies of China's water policy as it is applied to various international rivers and neighbouring countries: these range, geographically, from Central (Kazakhstan) to South Asia (Bangladesh and India), South-East Asia (ASEAN countries) and North-East Asia (Mongolia and Russia). In each of these chapters, the issue of insecurity over water basins is introduced against the context of foreign relations between China and the relevant neighbouring riparian state. In the unstable political conditions in Asia, these river basins represent specific situations where there exist complex political relations. The role China plays over major international rivers is then examined by assessing the causes and intensity of conflicts rooted in issues around how water resources are managed, the performance of relevant agreements and the outcomes of water diplomacy.

Chapter 9 examines how China's differing water relations are affected by contextual factors such as the magnitude of any given water management problem and the existence of conventional bilateral security concerns. In this way, each of the international rivers covered by the case studies will be compared in terms of the main dimensions of water politics, including the varying extent of political tensions that are displayed in Asia's international waters, and the differing outcomes of cooperative agreements where there is potential for the incitement of conflicts and other threats to regional stability. It provides a critical analysis of the effectiveness of different cooperative initiatives over water disputes between China and neighbouring states. Part of this comparative analysis is to investigate how different interests that impact on China's development of water diplomacy have been articulated.

China's water diplomacy has a versatile and changing character. Along with its water diplomacy, China is likely to be willing to commit to developing trust and collaboration over shared water management. Chapter 10 further elaborates on why China's water diplomacy, despite that temporary successes may have been achieved, is likely to fail in the long run. The chapter concludes by providing policy recommendations for how to best secure benefits from the sharing of international rivers in a peaceful and sustainable way.

## Notes

1 Personal communication with Russia diplomat.
2 These include the UN Convention on the Law of Non-Navigational Use of International Watercourses (1997); the UNECE Convention on the Protection and Use of Transboundary Watercourses and International Lakes (1992); and the Helsinki Rules (1966).
3 If rivers are boundaries between nations, more controversies are likely to occur as countries need to negotiate on the use and regulation of watercourse (Donaldson, 2009).
4 These are agreement on the Protection and Utilization of Border Waters Between China and Mongolia (1994); Cooperation Agreement on the Utilization and Protection of Transboundary Waters Between China and Kazakhstan (2001); Agreement on Reasonable Utilization and Protection of Transboundary Waters between China and Russia (2008).
5 For example, on 23 October 2013, the Border Defence Cooperation Agreement (BDCA) was signed between China and India.
6 'Joint Statement between the India and China during Prime Minister's visit to China', Press Information Bureau, Government of India, Prime Minister's Office, 15 May 2015. Available at: http://pib.nic.in/newsite/PrintRelease.aspx?relid=121755 (accessed 9 August 2015).
7 Guowuyuan Bangongting [State Council Office], Guowuyuan guanyu shixing zuiyange shuiziyuan guanli zhidu de yijian [Opinion of the State Council Regarding Implementation of the Most Stringent Water Resource Management System], 12 January 2012. Available at: www.gov.cn/zwgk/2012-02/16/content_2067664.htm (accessed 9 August 2015).
8 See Xi's speech on China's security strategy. Available at: http://china.org.cn/chinese/2014-06/03/content_32561159_2.htm (accessed 1 May 2017).
9 See UN Convention on the Protection and Use of Transboundary Watercourses and International Lakes. Available at: www.unece.org/fileadmin/DAM/env/documents/2013/wat/ECE_MP.WAT_41.pdf (accessed 9 August 2015).
10 China is one among three countries (along with Turkey and Burundi) that are against the adoption of the convention at the UN General Assembly.
11 United Nations Watercourses Convention 'United Nations Watercourses Convention enters into force', 18 August 2014. Available at: www.unwatercoursesconvention.org/news/united-nations-watercourses-convention-enters-into-force/ (accessed 9 August 2015).
12 See UN Watercourse Convention. Available at: www.unwatercoursesconvention.org/importance/the-relationship-with-the-unece-water-convention/ (accessed 21 November 2013).
13 See 1997 UN Convention on the Law of the Non-Navigational Uses of International Watercourses. Available at: http://internationalwaterlaw.org/documents/intldocs/UN_Watercourses_Convention-English.pdf (accessed 9 August 2015).

## References

Bandyopadhyay, J. and Ghosh, N. 2009. Holistic engineering and hydro-diplomacy in the Ganges-Brahmaputra-Meghna Basin, *Economic and Political Weekly*, 44(45): 50–60.

Beck, A. 2015. The UN Peacebuilding Commission and the potential of water in post-conflict development, governance and reconciliation, *Water International*, 40(2): 215–230.

Bernauer, T. and Kalbhenn, A. 2010. The politics of international freshwater resources, in R.A. Denemark, ed., *The International Studies Encyclopedia*, Malden, MA: Wiley-Blackwell, pp. 5800–5821.

Biba, S. 2014. Desecuritization in China's behavior towards its transboundary rivers: The Mekong River, the Brahmaputra River, and the Irtysh and Ili Rivers, *Journal of Contemporary China*, 23(85): 21–43.

Biswas, A. 2008. Management of transboundary waters: An overview, in O. Varis, C. Tortajada and A.K. Biswas, eds, *Management of Transboundary Rivers and Lakes*, Berlin: Springer, pp. 1–20.

Biswas, A. 2011. Cooperation or conflict in transboundary water management: Case study of South Asia, *Hydrological Sciences Journal*, 56(4): 662–670.

Breslin, S. 2015. Debating human security in China: Towards discursive power? *Journal of Contemporary Asia*, 45(2): 243–265.

Brochmann, M. and Gleditsch, N. 2006. Shared rivers and international cooperation. Paper presented at the Forty-Seventh Annual Convention of the International Studies Association, San Diego, CA, 22–26 March and the workshop 'Polarization and conflict', Nicosia, Cyprus, 26–29 April.

Brochmann, M. and Gleditsch, N. 2012. Shared rivers and conflict: A reconsideration, *Political Geography*, 31: 519–527.

Brochmann, M. and Hensel, P. 2009. Peaceful management of international river claims, *International Negotiation*, 14(2): 391–416.

Butts, H. 1997.The strategic importance of water, *Parameters*, 28(1): 65–83.

Caponera, D. 1987. The role of customary international water law, in M. Ali, A. Khan and G. Radosevich, eds, *Water Resources Policy for Asia*, London: Taylor and Francis.

Carius, A. 2006. Environmental cooperation as an instrument of crisis prevention and peace-building, *International Studies Perspectives*, 9(1): 1–21.

Chellaney, B. 2011. *Water: Asia's New Battleground*, Washington, DC: Georgetown University Press.

Chen, H., Clarke, A. and Wouters, P. 2013. Exploring China's transboundary water treaty practice through the prism of the UN Watercourses Convention, *Water International*, 38(2): 217–230.

Chowdhury, N.T. 2010. Water management in Bangladesh: An analytical review, *Water Policy*, 12(1): 32–51.

Collier, P. 2007. *The Bottom Billion: Why the Poorest Countries Are Failing and What Can Be Done About It*, Oxford: Oxford University Press.

Dietz, T., Ostrom, E. and Stern, P.C. 2003. The struggle to govern the commons, *Science*, 302(5652): 1907–1912.

Donaldson, J.W. 2009. Paradox of the moving boundary: Legal heredity of river accretion and avulsion, *Water Alternatives*, 4(2): 155–170.

Feng, Y., He, D. and Wang, W. 2015. Identifying China's transboundary water risks and vulnerabilities: A multidisciplinary analysis using hydrological data and legal/institutional settings, *Water International*, 40(2): 328–341.

Gain, A. and Schwab, M. 2012. An assessment of water governance trends: The case of Bangladesh, *Water Policy*, 14(5): 821–840.

Gerlak, A.K. and Grant, K.A. 2009. The correlates of cooperative institutions for international rivers, in T.J. Volgy, Z. Šabič, P. Roter and A.K. Gerlak, eds, *Mapping the New World Order*, New York: Wiley.

Gilley, B. 2013. *The Right to Rule: How States Win and Lose Legitimacy*, New York: Columbia University Press.

Gleick, P. 1990. Environment, resources, and international security and politics, in E. Arnett, ed., *Science and International Security: Responding to a Changing World*, Washington, DC: American Association for the Advancement of Science Press, pp. 501–523.

Goldstein, J. 1992. A conflict-cooperation scale for WEIS events data. *The Journal of Conflict Resolution*, 36(2): 369–385.

Grainger, S. and Conway, D. 2014. Climate change and international river boundaries: Fixed points in shifting sands, *Wiley Interdisciplinary Reviews: Climate Change*, 5(6): 835–848.

Haftendorn, H. 2000.Water and international conflict. *Third World Quarterly*, 21(1): 51–68.

He, D., Liu, C., Feng, Y., Hu, J., Ji, X. and Li, Y. 2014. Progress and perspective of international river researches in China [Zhongguo guoji heliu yanjiu jinzhan yu zhanwang], *Acta Geographica Sinica*, 69(9): 1284–1294.

Hensengerth, O. 2009. *Regionalism in China–Vietnam Relations: Institution-Building in the Greater Mekong Subregion*, Abingdon: Routledge.

Hill, D.P. 2015. Where hawks dwell on water and bankers build power poles: Transboundary waters, environmental security and the frontiers of neoliberalism, *Strategic Analysis*, 39(6): 729–743.

Ho, S. 2013. River politics: China's policies in the Mekong and Brahmaputra in comparative perspective, *Journal of Contemporary China*, 23(85): 1–20.

Homer-Dixon, T. 1994. Environmental scarcities and violent conflict: Evidence from cases, *International Security*, 19(1): 5–40.

Homer-Dixon, T. 1999. *Environment, Security and Violence*, Princeton, NJ: Princeton University Press.

ICA (Intelligence Community Assessment). 2012. *Global Water Security*. Available at: http://fas.org/irp/nic/water.pdf (accessed 10 July 2017).

Jahiel, A.R. 1998. The organization of environmental protection in China, *The China Quarterly*, 156: 757–787.

Jia, S. 2013. Ruhe fangzhi 'Santiao Hongxian' bian lv [How to prevent the Three Red Lines from turning green], in China Water Resources Association Water Resources Committee and Zhengzhou University Water Resources and Environment Institute, eds, *2012nian nianhui ji xueshu yantaohui lunwenji* [*Proceedings of the 2012 Annual Meeting and Symposium*], Henan: The Yellow River Water Conservancy Press.

Jia, S., Lv, A., Han, Y., Long, Q., Zhu, W. and Yan, H. 2014. *Water Resources Security Report of China*, Beijing: Science Press.

Kattelus, M., Kummu, M., Keskinen, M., Salmivaara, A. and Varis, O. 2015. China's southbound transboundary river basins: A case of asymmetry. *Water International*, 40(1): 113–138.

Khan, T.A. 1996. Management and sharing of the Ganges, *Natural Resources Journal*, 36(3): 455–479. Available at: http://digitalrepository.unm.edu/cgi/viewcontent.cgi?article=1780&context=nrj (accessed 10 July 2017).

Li, X. 2015. Applying offensive realism to the rise of China: Structural incentives and Chinese diplomacy toward the neighboring states, *International Relations of the Asia-Pacific*, 16(2): 241–271.

Liebman, A. 2005. Trickle-down hegemony? China's 'peaceful rise' and dam building on the Mekong, *Contemporary Southeast Asia: A Journal of International and Strategic Affairs*, 27(2): 281–304.

Liu, H., Geng, L., Zhong, H. and Gu, Y. 2006. On speeding up water resources development of international rivers of China [*Guanyu jiakuai woguo guoji heliu shuiziyuan kaifa liyong de sikao*], *Yangtze River*, 7: 32–33.

Liu, J., Zang, C., Tian, S., Liu, J., Yang, H., Jia, S., You, L., Liu, B. and Zhang, M. 2013. Water conservancy projects in China: Achievements, challenges and way forward, *Global Environmental Change*, 23(3): 633–643.

Lowi, M.R. 1993. *Water and Power: The Politics of a Scarce Resource in the Jordan River Basin*, Cambridge: Cambridge University Press.

Ma, X. and Ortolano, L. 2000. *Environmental Regulation in China: Institutions, Enforcement and Compliance*, Lanham, MD: Rowman & Littlefield.

Magee, D. 2012. The dragon upstream: China's role in Lancang-Mekong development. In J. Öjendal, S. Hansson and S. Hellberg, eds, *Politics and Development in a Transboundary Watershed*, New York: Springer, pp. 171–193.

Mason, M. 2008. The governance of transnational environmental harm: Addressing new modes of accountability/responsibility. *Global Environmental Politics*, 8(3): 8–24.

McIntyre, O. 2014. The protection of freshwater ecosystems revisited: Towards a common understanding of the 'ecosystems approach' to the protection of transboundary water resources, *Review of European, Comparative and International Environmental Law*, 23(1): 88–95.

Moore, S.M. 2014. Hydropolitics and inter-jurisdictional relationships in China: The pursuit of localized preferences in a centralized system, *The China Quarterly*, 219: 760–80.

Moynihan, R. and Magsig, B.-O. 2014. The rising role of regional approaches in international water law: Lessons from the UNECE water regime and Himalayan Asia for strengthening transboundary water cooperation, *Review of European, Comparative & International Environmental Law*, 23(1): 43–58.

Mushkat, R. 2010. Compliance with international environmental regimes: Chinese lessons, *William & Mary Environmental Law and Policy Review*, 34(2): 493–542.

Ovodenko, A. 2014. Regional water cooperation: Creating incentives for integrated management, *Journal of Conflict Resolution*, 60(6): 1071–1098.

Raadgever, G.T., Mostert, E., Kranz, N., Interwies, E. and Timmerman, J.G. 2008. Assessing management regimes in transboundary river basins: Do they support adaptive management? *Ecology and Society*, 13(1): art. 14. Available at: www.ecologyand society.org/vol13/iss1/art14/ (accessed 10 July 2017).

Salman, S.M.A. 2007. The United Nations Watercourses Convention ten years later: Why has its entry into force proven difficult? *Water International*, 32(1): 1–15.

Scheumann, W. and Hensengerth, O. 2014. *Evolution of Dam Policies: Current Practices and the State of the Debate*, Berlin: Springer.

Schmeier, S., Gerlak, A.K. and Blumstein, S. 2016. Clearing the muddy waters of shared watercourses governance: Conceptualizing international River Basin Organizations, *International Environmental Agreements: Politics, Law and Economics*, 16(4): 597–619.

Shapiro, J. 2012. *China's Environmental Challenges*, Cambridge: Polity Press.

Shen, D.J. 2014. Post-1980 water policy in China, *International Journal of Water Resources Development*, 30(4): 714–727.

Silveira, A.R. da and Richards, K.S. 2013. The link between polycentrism and adaptive capacity in river basin governance systems: Insights from the River Rhine and the Zhujiang (Pearl River) basin, *Annals of the Association of American Geographers*, 103(2): 319–329.

Sonnenfeld, D.A. and Mol, A.P.J. 2002. Globalization and the transformation of environmental governance: An introduction, *American Behavioural Scientist*, 45(9): 1318–1339.

Sood, A. and Mathukumalli, B.K.P. 2011. Managing international river basins: Reviewing India–Bangladesh transboundary water issues, *International Journal of River Basin Management*, 9(1): 43–52.

Tarlock, A.D. 2014. Mexico and the United States assume a legal duty to provide Colorado River delta restoration flows: An important international environmental and water law precedent, *Review of European, Comparative and International Environmental Law*, 23(1): 76–87.

Tir, J. and Ackerman, J.T. 2009. Politics of formalized river cooperation, *Journal of Peace Research*, 46(5): 623–640.

Tolba, M.K. with Rummel-Bulska, I. 1998. *Global Environmental Diplomacy: Negotiating Environmental Agreements, 1973–1992*, Cambridge, MA: MIT Press.

UNEP (United Nations Environment Programme). 2012. *Status Report on the Application of Integrated Approaches to Water Resources Management*, Nairobi: UNEP. Available at: www.un.org/waterforlifedecade/pdf/un_water_status_report_2012.pdf (accessed 17 August 2017).

Uprety, K. and Salman, S.M.A. 2011. Legal aspects of sharing and management of transboundary water in South Asia: Preventing conflicts and promoting cooperation, *Hydrological Sciences Journal*, 56(4): 641–661.

Van Rooij, B. 2006. *Regulating Land and Pollution in China: Lawmaking, Compliance, and Enforcement – Theory and Cases*, Leiden: Leiden University Press.

Varis, O., Kummu, M. and Salmivaara, A. 2012. Ten major rivers in monsoon Asia-Pacific: An assessment of vulnerability, *Applied Geography*, 32(2): 441–454.

Warner, J. and Mirumachi, N. 2008. Co-existing conflict and cooperation in transboundary waters. Paper presented at the international studies association annual conference. San Francisco, 26–30 March.

Wolf, A.T. 1999. Criteria for equitable allocations: the heart of international water conflict. *Natural Resources Forum*, 23(1): 3–30.

Wolf, A.T., Kramer, A., Carius, A. and Dabelko, G.D. 2005. Managing water conflict and cooperation, in A.T. Wolf, ed., *State of the World: Redefining Global Security*, Washington, DC: Worldwatch Institute.

Wolf, A.T., Yoffe, S.B. and Giordano, M. 2003. International waters: Identifying basins at risk, *Water Policy*, 5(1): 29–60.

Wolf, R. 2011. Respect and disrespect in international politics: The significance of status recognition, *International Theory*, 3(1): 105–142.

Xiao, R. 2007. Yanjiu he lijie Zhongguo de guoji zeren [Studies and understandings of China's international responsibility], *Shehui kexue* [*Journal of Social Sciences*], 12: 27.

Xie, L. 2009. *Environmental Activism in China*, Abingdon: Routledge.

Yeophantong, P. 2013. Governing the world: China's evolving conceptions of responsibility, *The Chinese Journal of International Politics*, 6(4): 329–364.

Yoffe, S.B., Wolf, A.T. and Giordano, M. 2003. Managing water conflict and cooperation, in Worldwatch Institute, ed., *Redefining Global Security: The State of the World Report*, Washington, DC: Worldwatch Institute.

Zawahri, N.A. and Gerlak, A.K. 2009. Navigating international river disputes to avert conflict, *International Negotiation*, 14(2): 211–227.

Zeitoun, M. and Mirumachi, N. 2008. Transboundary water interaction I: Reconsidering conflict and cooperation. *International Environmental Agreements*, 8(4): 297–316.

Zeng, J., Xiao, Y. and Breslin, S. 2015. Securing China's core interests: The state of the debate in China, *International Affairs*, 91(2): 245–266.

Zha, D. 2015. A political ecology of hydropower development in China, in W. Jisi ed., *Review of China's International Strategies*, Beijing: Center for International and Strategic Studies, Peking University.

Zheng, Y. 2001. Introduction, in Environment and Development Research Institute, ed., *China Environment and Development Review*, Beijing: Social Sciences Documentation Publishing House.

# 2 The governance of water resources in China

## Introduction: China's perception of water as a multipurpose resource

Water, an essential resource for human well-being, is regarded as a key source of economic growth in China. China has a 4,000-year history of this approach to water management, focusing on agriculture and flood control as the main concerns. In ancient times, a notion of human–nature coordination was advocated. The management of the Yellow River, in particular, was an essential task for many dynasties. Some even suggest that China's political system was built upon the administrative system which governs water-related affairs (Wei, 2011). In Mao's time, management and control were a strong focus of China's approach to the use of water. This is also referred to as *Shuili*, which means benefits from water resources and avoiding the difficulties that water disasters could bring. Since 1949, laws and legal documents have confirmed that water is of instrumental value in facilitating economic growth and societal needs in China. For instance, in the Chinese Water Law (see Table 2.1), a founding principle that provides guidance in the management of water in the country, there is an ambiguous use of 'water resource' as an alternative word for water. In this legal document, the Chinese government encourages and supports all undertakings for the development and utilization of water resources. And the uses of water are primarily perceived as of economic value. The exploitation of water resources for hydropower, transport navigation and other for-profit uses is highly encouraged.

Since the early 2000s, however, there remains a lack of consideration of the intrinsic values of rivers, independent of water construction projects. For instance, there has been an increasing concern shown towards the conservation of rivers. The Yellow River is one of the most debated. It has been seen as nurturing the ancient Chinese civilization and a cultural understanding has highlighted the environmental importance of its river basins. Chinese domestic critics have highlighted this weakness in China's prioritizing its water management and governance only for economic purposes. Such a prioritizing has also already been shown to have limitations in practice (Cai, 2007).

Therefore, the Chinese perception of the sustainable development of water is understood differently from the way it is understood by western counterparts.

*Table 2.1* Legal documents on water management in China

| *Name of the law* | *Purpose* |
|---|---|
| Law on the Prevention and Control of Water Pollution (*shuiwuran fangzhi fa*) (1984, 1996 and 2002, amended 2008) | Provisions for the enforcement and clarification of responsibilities in water pollution control and the conservation of water ecology |
| Water Law (*shuifa*) (established in 1988, revised in 2002) | Core document, providing principles in the planning and management of water resources, and emphasizing water saving and sustainable development |
| Environmental Protection Law (*huanjing baohu fa*) (1989, revised in 2014) | Lays the foundations for policies and principles for the protection of the natural environment and the management of natural resources |
| Water and Soil Conservation (*shuitu baochi fa*) (1991) | Regulates water management and the protection of water ecology in major construction projects |
| The Prevention and Control of Environmental Pollution by Solid Waste (*guti feiwu wuran huanjing fangzhi fa*) (1996) | Aims to control the pollution from solid wastes. It touches upon the prevention of pollution of rivers and river basins |
| Flood Prevention and Control Law (1997) (*fanghong fa*) | Provides guidance on the control of floods and the prevention of damage. It regulates the planning and management of river basins from the perspective of disaster prevention |
| Environmental Impact Assessment Law (*huanjing yingxiang pingjia fa*) (2002) | Touches upon the prevention of pollution and protecting the ecology of river basins |

Note
All measures have been prescribed by the National People's Congress.

The Chinese government has demonstrated a clear anthropocentric view that strongly advocates human needs and interests. The government's attitude towards the non-human world remains subordinate to the consideration of human society's interests (Shapiro, 2012). China focuses on the 'sustainable uses' of water as multiple uses for economic purposes (Wang and Wang, 2012).

## Security of China's water uses

China's water uses and management are of significant impact to the global economy. On the one hand, because of the sheer size of China, its commodities importation and the requirement to import food as a result of water shortages in the country can have a large impact on international markets. On the other hand, China is a country that is the source of many international rivers. Its domestic water uses can have strong global impacts. Indeed, more concerns have arisen about the country's internal water uses including its water diversion projects. For instance, water diversions from China's international river basins will affect the water supplies of nearby countries such as Vietnam, Laos, Cambodia, India, Thailand and Bangladesh.

China enjoys an overall water resource of 2,780 km³ and has demonstrated that on a national scale, the overall volume of water supply from the country is sufficient to support its economic and industrial development (see Table 2.1 and Figure 2.1). Various scientific studies have been conducted to understand China's water uses and how the country responds to increasing water management challenges. It is assumed that the national government has made plans to allocate water volumes for ecological uses with the remainder being made available for human uses. We adopt the study of Jia *et al.* (2014). On national scale, the country enjoys a potential water supply of 1,000 km,[3] which takes into consideration of the amount of water that is cycled back from nature. According to the Three Red Lines policy (TRL) in 2011, the latest water policy, the overall national water use is controlled at 700 km³ for societal consumption. Considering the amount of water that is cycled back from nature, the overall national consumption is allowed to reach 810 km³,[1] which falls well under the estimated level of water supply at national level.

China has also begun to experience a transformation of its economic structure, leading to significant decreases in its industrial pollution. It is noteworthy that the overall water consumption in China has started to decline. According to Jia and Zhang (2000), it is predicted that China's peak water use will not

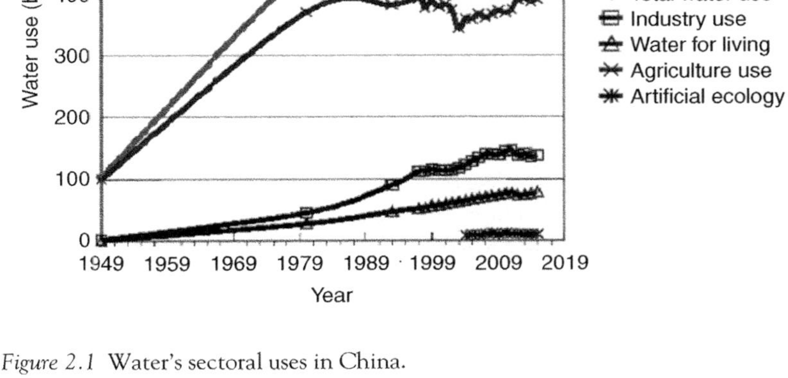


*Figure 2.1* Water's sectoral uses in China.

Source: Jia *et al.* (2014).

exceed 650 km$^3$ per year. This has been proven by figures in 2013, when the country's peak water use reached 618.3 km$^3$, and then decreased to 609.5 km$^3$ and 618.0 km$^3$ in 2014 and 2015, respectively (MWR, 2014, 2015, 2016). Industries that are more likely to have discharges, such as steel, manufacturing, and printing and dyeing, are being replaced by less-polluting industries such as electronics or finance and services. Less discharge indicates that the country's total water use has also declined. It is further argued that a Kuznets curve can be found in water uses which show that a reduction in water use can be achieved along with a country's economic growth (Jia *et al.*, 2006).

China has rich water resources, which include 40 major transboundary water courses shared with neighbouring countries in the north-west, north-east and south-west. They are Afghanistan, Bangladesh, Bhutan, Cambodia, India, Kazakhstan, Kyrgyzstan, Laos, Mongolia, Myanmar, Nepal, North Korea, Pakistan, Russia, Tajikistan and Vietnam. Among China's rivers, 15 are major rivers. Twelve of the rivers originate in China, and in some cases there are multiple river courses shared with countries that include Russia, Mongolia and Myanmar. Yan (forthcoming) indicate that among ten transboundary river basins studied, for about a half of the basins, water resources within China's territory accounts for over a quarter of the basin's water resources (see Table 2.2). This is the case for the basins of the Mekong, the Nu, the Red, the Amur, the Ili and the Irtysh Rivers. On the Indus, the Ganges, the Ob and the Yarlung Zangbo Rivers, China does not possess the largest proportion of the basin's water resources.

However, China is faced with several of its shared water basins being increasingly overexploited, most of which accumulate in the Amur River in the north-east and the north-west, including the Ob, Ili, Tarim and Irtysh Rivers. On the Amur River, China has shown a high degree of water exploitation, with its ratio of water resources to the total basin reaching 40.5 per cent. Water shortage is the main cause of this, as China and its neighbouring states, such as Kazakhstan (shared for Ili and Irtysh Rivers) and Kyrgyzstan (shared with Tarim River) are located in desert areas where natural resources are limited.

## Water needs and pressure

China's strategy towards water management has evolved dramatically. In ancient China, water use was primarily focused on agriculture. The country's transition to an industrialized nation saw dramatic changes in its water usage, including an expansion in industrial uses and the development of various water investment projects. Around 1978, economic reform accelerated and China's rapid industrialization and urbanization resulted in a growing demand for water in agricultural and non-agricultural sectors, such as in cities, industry and township enterprises. Since China's economic reform in the 1980s, a strategy known as the 'primary stage of socialism' (Zhao, 1987; Shen, 2014) was adopted, which advocated and promoted development of the economic sector. Pressure was exerted in relation to water management, with water policy being made to support the development of the national economy and society as a whole, not

*Table 2.2* Water security of river basins

| *Basin name* | *Water resources ($10^9 m^3$)* | *Water resources of China ($10^9 m^3$)* | *Ratio of China to the total basin (%)* | *Withdrawal of China ($10^9 m^3$)* | *Ratio of China's withdrawal to the total basin withdrawal (%)* | *Ratio of China's uses of water resources to the total basin (%)* |
|---|---|---|---|---|---|---|
| Indus River | 1,895.6 | 64.4 | 3.3 | 0.1 | 3.9 | 0.1 |
| Ganges River | 4,542.4 | 323.5 | 3.0 | 0.8 | 2.4 | 0.6 |
| Mekong River | 6,929.1 | 749.4 | 26.1 | 28.0 | 9.0 | 3.7 |
| Ob River | 3,850.0 | 100.8 | 2.6 | 15.4 | 6.8 | 15.3 |
| Red River | 1,231.0 | 450.0 | 36.6 | 28.3 | 21.0 | 6.3 |
| Nujiang/Saerwenjiang | 1,739.4 | 720.9 | 39.1 | 13.1 | 33.9 | 1.8 |
| Heilong (Amur) River | 3,465.0 | 1,122.3 | 32.4 | 454.2 | 95.8 | 40.5 |
| Yarlang Zangbo River | 9,425.4 | 3,334.3 | 13.1 | 24.7 | 6.5 | 0.7 |
| Ili River | 241.1 | 156.9 | 65.1 | 56.8 | 61.3 | 36.2 |
| Irtysh River | 377.6 | 100.8 | 26.7 | 15.4 | 30.9 | 15.3 |

Source: Yan (forthcoming).

simply focusing on agriculture, as in the past. This required the economic sector to accommodate industrial uses, and to coordinate increased water demands for use in hydropower projects, with water and soil conservation and water sources protection (Qian, 1985). At the same time, with increasing concerns about urban water shortages in Northern China, demands have grown for better-integrated management of water resources (Qian, 1985).

In the 1990s, the water management situation in China worsened. With the growth of cities and industry onto farmland, there has been a shift in the country's grain basket from the wet but mountainous south to the flat, dry north, and with it a growing reliance on receding groundwater resources. Frequent water-related disasters and incidents of serious water pollution have occurred. The demands of a rapidly expanding urban population and industry have added to the burdens on the water supply and in many places have resulted in the appropriation of water from rural areas. In recent years, total water use has begun to rise and there have been growing concerns about water quantity and quality issues, including the pollution of both surface and underground water resources. The concerns expressed have now reached alarming levels. This calls for better water-use efficiency measures.

## Legal framework and water management approaches

China has not yet developed a clear set of rules to regulate river basin management. On a number of occasions, relevant regulations have been incorporated into general water-related legislation. The promulgation of the Water Law in 1988 marked a shift in establishing a legislative system to regulate China's water resources. It has also explicitly promoted the development of integrated water resources management, giving stronger attention to water quality and ecology, which has strong implications for human security (Huang, 2004). Apart from this national law, most legislation takes the form of ministerial laws that strongly reflect ministerial interests.

The Water Law is one of the founding laws in China which regulates the use of water resources.[2] It has prescribed regulatory control principles and guidance for the management of water resources. In particular, it provides guiding principles to resolve water conflicts; for instance the allocation of water supplies between users of water basins (which comes under administrative jurisdiction) is identified as one common type of water dispute, particularly among upstream–downstream users (Chu *et al.*, 2014). In the following sections, Chinese government's principles in exercising water rights and key concerns in water uses are analysed.

### *Principles in accessing water*

The Chinese government is aware that it is in need of clarification of how to best manage abstraction of its water resources. Theoretically and in practice, China is experiencing heated debate on what 'a water right' means. Although the country has developed a recognition of rights to access water resources, contention exists on what legitimate users are and what entitlements they have. Indeed, studies have shown that in practice the law has raised ambiguity and has caused conflicts among water users (Shan, 2016).

The conceptualization of a water right differs among stakeholders. It is also sometimes referred to as a water property right. It is suggested that the primary concern for regulating water rights is the allocation of and the access to water resources (Wang, 2000; Zhang and Bai, 2004). This legal framework prescribes the government's legitimacy as an authority and in undertaking administrative activities and arrangements. According to Article 9 in the Constitution and Article 3 in the Water Law, water resources, including both surface and underground water, are in the possession of the state. However, water rights are also open to individual users. A permit system is adopted for users to rightfully appropriate water resources.[3] This principle, then, emphasizes the timing of water users in their appropriation of water resources such that those who are the earliest to appropriate water are granted the right to that water. However, ambiguity has occurred in users' acquiring water rights and injustice is likely to occur. For one, it has become a source of the 'first served, first use principle' (Shan, 2016). Those who are more resourceful have a better chance to be entitled with water rights. Similarly, others advocate the principle of distance. That is, those who are located closest to the water they wish to acquire are endowed with the water right (Li *et al.*, 2011), leaving disadvantaged those who are located afar.

Another ambiguity lies in its implication that the right to appropriate water depends on the purposes for which the water resources are to be used. It is stated in Article 4 of the Water Law that the uses of water resources are to be well-coordinated to reach a balance between resource management and human development. As human development and well-being is only broadly understood and the extent to which human development depends on water resources varies, the concept of a water right is therefore represented in a weak fashion. More specifically, three broad areas are distinguished for human use, categorized as gaining economic profits (*Shengchan*), freshwater and sanitation (*Shenghuo*), and ecological protection (*Shengtai*). It is implied that equity in water rights means that basic human health considerations are to be included when implementing the primary concern on water resources. Hence, priority should be provided to households and human well-being, before the consideration of other uses, such as agriculture, industry, the preservation of water ecology, and watercourse transport navigation. However, such regulations imply inconsistency with the requirement of sustainable development, which is an underlying principle in water resource management.[4] The attention to water ecology seems unbalanced with seeking economic interests or satisfying human well-being from water resources (Cai, 2007). Without explicit accounts, it remains ambiguous if and how this principle of sustainable development applies.

Additionally, it is also argued that the Water Law does not clearly rank the purposes that water resources are to serve. Apart from the blurred wordings, no details are provided as to how this prioritizing is to be gauged. An increasing number of disputes have arisen between irrigation and household users. Hence, there is a pressing need to clarify the principles that facilitate the fair allocation of and access to water resources (Shan, 2016).

### *Water management areas*

China's water management has undergone dramatic reforms that have touched upon the combined policy areas of water planning and resource management. These reforms have brought about a coordinated approach to creating harmony between the needs of human society and nature. As a result, significant changes have occurred that impact on the sustainable development of river basins. China's management of water supplies previously relied on unlimited abstractions of water. This approach relied on investments in water projects to continuously raise supplies of fresh water. More recently, China's water resource management has placed a high significance on flood control, and the construction and operation of associated infrastructure has been a key issue for the central government (Chi, 1963).

Along with recognizing the country's pressing water needs, China has implemented the strategy of 'resource-oriented water resources development' (Wang, 1999). Apart from a key focus on quantity control, an expansion in water management initiatives has incorporated more concern for water resource conservation, protection, irrigation and water supply (Liu and Speed, 2009). A deepened understanding of anthropogenic impacts on water resources has developed. This has led Chinese policymakers to shift their attention away from developing new water sources towards a greater appreciation for resource management, environmental effects, and demand management including waste reduction (improved efficiency).

Reflected in China's legislation on environmental protection and water resource management, the protection of water ecology has been incorporated more closely into the planning and management of the country's water resources. Over the past three decades, most national laws or newly revised legal documents have incorporated the protection of water quality as a guiding principle (see Table 2.1). For instance, the Water Law (1988) clarifies ways to promote the development of integrated management of China's water resources, placing a higher level of attention on the areas of water quality and ecology (Huang, 2004). According to the Opinions on Implementing the Development of Five Systems for the Water Sector released by the Ministry of Water Resources (MWR) in 1993, a better and more effective water service is to be developed, focusing on demand accumulated by the whole of society (Wang, 2000). The 2002 revision of the Water Law established the right for society to access safe water, and elevated the importance of conservation and pollution control as water policy objectives (Xie *et al.*, 2009: 28). In particular, ecological protection of water resources has become a key focus in water resource management. For example, the 2014 revision of the 1989 Environmental Protection Law has strengthened the requirement for protecting the ecological functions of sensitive and fragile environments (Spijkes *et al.*, 2014). This law requires the state to provide guidance to the local governments of beneficiary areas and ecological protected areas on how ecological compensation can be conducted. Both public consultation and market rules are encouraged to ensure such a mechanism works fairly and efficiently. This has resulted in an increased fiscal transfer to areas of ecological protection (Spijkes *et al.*, 2014).[5]

More systematic development of policy has recently occurred in China (Dai, 2015). In 2011, Policy Document No. 1 (2011) was promulgated to establish a holistic water resources management strategy. This policy incorporates two other important policy documents, the Strictest Water Resources Management System (SWRMS) (2012) and the Three Red Lines (TRL) policy. Through these policies, the government has established a water resources management strategy which has strong implications for promoting sustainable development. The strategy addresses the issues of water allocation, conservation and protection, and takes full account of measures to control both water quantity and quality. By setting out quantitative targets, it aims to promote integrated resource management of various water bodies, including river basins (Zhang *et al.*, 2016).

By firmly relying on jurisdiction as a basis, the policy document aims to promote the decentralization of water management responsibilities to the local level including the county level and even county-level administrative units (Jia, 2012). It is believed that such arrangements will promote the maximization of profits that can be sought from water resources, while simultaneously protecting a healthy aquatic water ecology (Jia, 2012). A few of China's large river basins have also received attention resulting in specific regulatory institutions regarding water management. For example, in Xinjiang the local government has set out laws to regulate the management and protection of the Ili River basin.[6] This provincial law has outlined specific ecological functions of the river basin, indicating concerns that include the loss of biodiversity, water pollution and the degradation of grassland. On domestic transboundary rivers, heated debate has been seen over establishing laws for the management of the Yellow River basin (Cai, 2007).

For the first time, the importance of water management to national development has been stressed. Referred to as a 'national security' concern, the urgency of improving the management of water resources is highlighted, reflecting that China is faced with growing pressure in its uses of water. In government's policy document promulgated in 2011, it is states that

> Promoting the development of water reform, is not only important to agriculture and rural development, but also of significance to social, economic development; such reform is important … but also essential to the security of economy, ecology and the state.… Water reform should be prioritized on the political agenda of the Party and the state.[7]

## The planning and management of transboundary rivers

An increasing number of conflicts has arisen over water quantity allocations (Cai, 2007) and pollution control (Liu and Speed, 2009; Winalski, 2009; Hu and Cheng, 2013; Jiang, 2015). This section reviews policymaking and implementation in the planning and management of water resources. Emphasis is placed on analysing institutional arrangements adopted in the management of transboundary water basins.

### *Institutional arrangements for international river management*

China manages its river basins through complex institutional arrangements and legal regulations. Its administrative reforms have impacted on central–local intergovernmental relations, which has affected water management arrangements in the country. Since 1978, China has gradually transformed from a highly centralized system to a planned system that oversees economic decentralization and has created a federal financial system (Zheng, 2007). Central authorities depend on local governments for political support and financial contributions (Cheng and Hu, 2012). On water management, bargaining and negotiation has increased, supervision is weak and local pollution control is not strict (see also Jia *et al.*, 2016). Such localism has been another factor which impacts the implementation of national water policies. According to Jia *et al.* (2016), as a result, local users have extended rights in their appropriation of water.

Regulations set out in legal documents require that, apart from the central authorities, provincial and local governments have no entitlement to make decisions on water management. These government units are instead responsible for policy implementation. When viewing the management of transboundary water basins, much research has been undertaken from the perspective of central–local relations (Wang, 2008). Since China's economic reform in 1978, local authorities (provincial level and below) have acquired greater economic power and have formed a de facto federation (Zheng, 2007). Some even assert that fiscal centralization is a significant component of the economic opening-up policies reforms (Wang and Ma, 2014). Chien (2010) refers to China's case as 'asymmetric decentralization' – economic decentralization in the absence of political decentralization. This trend has also been seen in China's water governance. In a de facto federation such as China, increasing bargaining behaviours have been seen in central–local government negotiations, and are suggested as 'normalized' exercises in the decision-making processes (Zheng, 2007; Weingast, 2009). Such interactions play a significant role in the decision-making for river basin management, affecting the policy outcomes and outputs. The management of water, therefore, largely depends on the negotiations and bargaining between local and central authorities (Cheng and Hu, 2012; Jiang, 2015).

China's water management is mostly based on the country's administrative jurisdiction. China's bureaucracy, which features a unitary system, further complicates the process. It sees fragmented or *Tiao-Kuai* lines of authority (see Figure 2.2). In this system, line relationships represent administrative control, and governmental units have authority over each other. This authority only comes from the related agency functioning in the same line. The unit's link with other government units within the same jurisdiction is described as an area (*kuai*) relationship. Governmental units of the same jurisdiction do not have discretionary power over others at the same level. This results in a traditional administrative demarcation between government units that take charge of two interrelated issues in China's water resource management, being water resource developments and the control of water quality.

Overall, China's regulation and management of water is promoted by its top-down administration. The ownership of water is vested in the people that are represented by the State Council. It is the main actor in policymaking and its responsibilities include the formulation of water laws, policy on the planning

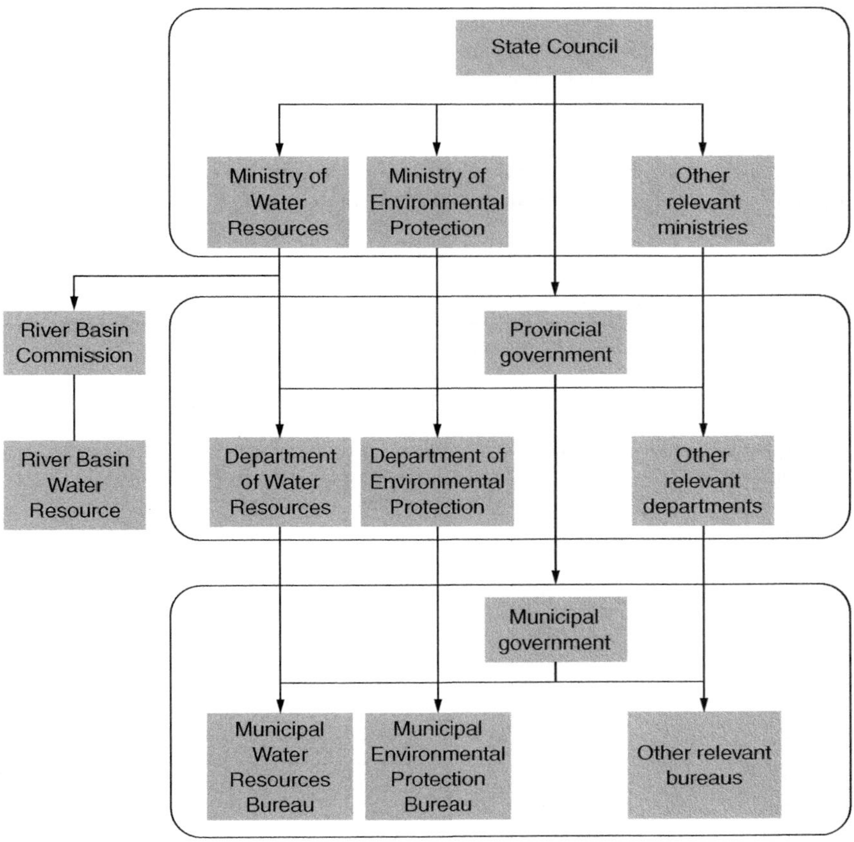


*Figure 2.2* China's institutional arrangements for water resource management.

and management of water resources, and decisions about water resource fees or taxes. According to institutional arrangements, the MWR and the Ministry of Environmental Protection (MEP) are the leading agents in water management. Since the 1980s, the State Council has given way to the MWR, which is now the lead government unit in China's overall water resource management. The MWR had been given overall authority over other ministries, including for the management of flood control and inter-basin transfer projects. In comparison, the MEP's authority has been enhanced since the middle of the first decade of the twenty-first century in line with the increasing pressure to better protect the natural environment. It is responsible for water quality control while the MWR is in charge of the allocation of water resources. Apart from these two main government units, a number of other relevant agencies have discretion on water management. These include the Ministry of Land and Resource, the National Development and Reform Commission (NDRC), and the Ministry of Transport and these agencies all have differing responsibilities (see Table 2.3). For instance, the Ministry of Agriculture and the Ministry of Transport place a strong focus on the use of water for irrigation, hydropower, municipal water supply and water transport.

Due to the lack of coordination and contradictory regulations, China's institutional arrangements have impeded water management, particularly for its domestic transboundary river basins. At the central government level, authorities are fragmented and power struggles exist among government units that represent different sectoral interests. This has been referred to as 'nine dragons

*Table 2.3* Central government units' responsibilities in China's water management

| *Unit* | *Remit* |
|---|---|
| Ministry of Water Resources | Responsibilities for water planning and management |
| National Development and Reform Commission | Responsible for investment planning and macroeconomic policies for water projects |
| Ministry of Land Resources | Responsible for the planning and management of groundwater resources |
| Ministry of Construction | Responsible for the construction and operation of water facilities, including water supply systems, water saving, urban flood control, water use planning and municipal wastewater facilities |
| Ministry of Health | Responsible for formulating and implementing policy to ensure the public's access to clean water |
| Ministry of Finance | Responsible for budgeting and investment |
| State-owned Assets Supervision and Administration Commission | Responsible for the management of state-owned enterprises that are in charge of water project constructions |

ruling the water' (九龙争水; Zhong, 2007). According to the Law of the People's Republic of China on Prevention and Control of Water Pollution, the MEP and its subordinate units at different levels are responsible for the regulation of water quality. The MEP is responsible for maintaining the Law of the People's Republic of China on Prevention and Control of Water Pollution which promulgates that water basin units are responsible for the protection of water quality. This is to say that both the administrative government units responsible for water quality as well as basin-scale government units are authorities over water resource management. Hence, the separation has impacted on the holistic management of water resources in China (Cai, 2007; Dai, 2015). The development, planning and implementation of water schemes within China's administrative boundaries cannot be implemented as an 'integrated watershed plan' because power struggles exist. Pollution regulation, which is a core area of MEP's work, it has shown a reluctance to adopt the collectively agreed standards. Instead, as a result of the power struggle between ministries, it has been keen to set up new standards according to its own regulatory measures (Jia, 2012).

### *Conflicts and tensions in water basin management*

China's complex institutional arrangements have shown weakness in its water governance. On water basin management, such situation has become more complicated, as there exists a large overlap in responsibilities between government units within and across scales (Pohlner, 2015). The bureaucracy created by the overarching legal document on water resource management, has led to multiple authorities with access to water uses and various interests uncoordinated (Pohlner, 2015). To some extent, the current institutions are feared to have contributed to conflicts that arise in the management of transboundary rivers (Wang *et al.*, 2007).

Vertical conflicts exist between Chinese central and local authorities on the management of transboundary river basins. At the regional and national level, the State Council decentralizes the water ownership right to local governments and individual users. The planning and management right remains with the national government which is responsible for decision-making on the planning and allocation of water resources. At the same time, users also have the right to manage water resources. This means that local governments are important units in implementing decisions. According to the Regulation on the Administration of the Licence for Water Drawing and the Levy of Water Resource Fees (2006), provincial governments are entitled to absolute discretion over water resource allocations and uses.[8]

Similarly, no clarification is provided as to how absolute authority and power is to be enforced on water arrangements, particularly for rivers that run through different neighbouring provinces (Jia *et al.*, 2016). The Water Law prescribes that the management of water basins is to be incorporated into the regional-level administration. Article 12 of the Water Law states that the 'centralized management is to be combined with multiple government levels and the different sectoral interests of government units that have responsibilities over water'. Tensions occur over domestic transboundary rivers between various provincial governments which adopt a range of competitive strategies to pursue their own localized interests at the expense of neighbouring jurisdictions. China has also set up River Basin Committees for each of its seven largest rivers. These committees are subordinate organizations of the MWR and have limited autonomy. They play only a weak role in coordinating different provinces in basin-based water management. As they are financially associated with provincial governments, these units are not autonomous actors; rather, they are subordinate to the provincial authorities of where they are located. According to the Regulation on Water Abstraction promulgated by the State Council (2006), provincial governments are entitled to absolute discretion on water resource allocations and uses. Hence, negotiation and renegotiation constantly occurs to temporarily settle disagreements (Yang *et al.*, 2016). And apart from various negotiations between provincial governments and central–local bargaining, no other actors are involved (Chu *et al.*, 2014).

### *Public participation in water governance*

China has shown a tendency to decentralize power to local authorities, with central authorities promoting basin-wide regional collaboration for the governance of transboundary rivers. This trend reflects a global pattern, whereby a decentralization of water management tasks is occurring as part of a broader paradigmatic shift in governing. It is generally agreed that this process is often perceived as empowering new local-level actors at the expense of central government agencies. However, much scholarly work has been developed on whether the notion of 'collaborative governance' can be immediately applied to China;[9] and whether such practices are occurring (Benson *et al.*, 2013).

Contentious views exist about the features and outcomes of collaborative governance in China's water management. China has seen decentralized water management policy experiments, such as in Northern China, including in Inner Mongolia, Hebei, Henan, Liaoning, Shanxi and Shaanxi Provinces. By comparing regional collaboration in Zhujiang delta with that of the Rhine in Europe, Silveira and Richards (2013) suggest that in the Chinese case, despite the political environment, local governments are strong in implementation, and they manage to develop collaboration at various other levels of government based on shared interests and goals for economic growth. Reviewing cases of dam constructions in transboundary water management, Moore (2014) has also observed that decentralization has the tendency to increase local authorities' power, and hence local governments are likely to prioritize their own interests in water management by ignoring other provinces' interests, ultimately impacting on the implementation of central policies. Similarly, a World Bank report indicates that ineffective implementation of water policies is a result of both insufficient transparency and local authorities deliberately resisting measures that could result in reductions of the limited tax revenue at their disposal (Xie *et al.*, 2009). Therefore, it has also been noticed that in China decentralization has created a multiple-centred water governance system that facilitates collaborative water governance.

Due to China's size and the uneven development across its regions, central–local governmental relations have been a continuous issue in the country. It is generally found that in regions where the gross domestic product (GDP) level is higher, local authorities exhibit varying degrees of commitment to implement water policies. Recent studies have shown that the central government's newly launched personnel management tool proves successful when local officials' performance is assessed based on water policy implementation (Dai, 2015). Hence, the localized policy instrument is suggested as a crucial factor in effective policy implementation (Dai, 2015).

The Chinese government has been called upon to focus more on environmental health in its water governance (Holdaway, 2010), with the intention to enhance public participation. Domestically, increasing societal concerns have been shown for water management, and growing social protests have become a source of anxiety for the central government. These have attributed to the Chinese government's giving more attention to problems such as polluting and hazardous industries, and food safety (Holdaway, 2010). Internally, with the increasing occurrence of water-pollution-related accidents and natural disasters, the negative environmental impacts on human health have been recognized (Zhang and Zhong, 2010). China is experiencing an increasing number of internal water disputes. From 2001 to 2005 over 40,000 disputes occurred between individuals or between different provinces. This has been regarded as a source of 'mass incident' (*quntixing shijian*) (Shan, 2016). The uneven distribution of water resources has meant that most disputes have arisen in the country's west where water scarcity is a much more serious issue than in the east (Shan, 2016). Similarly, more disputes are identified in large cities, where intensive water use exists (Liu and Speed, 2009). Such activity also occurs against the backdrop of more 'peaceful' action, notably the mobilization of public opinion via national and regional media campaigns as well as social media networks. As a result, the Chinese government has become cautious about causing mass incidents that are related to the environment. The promulgation of a series of environmental laws indicates that the central government has become aware of the seriousness of environmental issues for maintaining its rule. The National Emergency Public Incident Legislation (2006) provides regulation in response to public incidents related to natural disasters where an emergency response is required. This legislation stresses that international rivers have transboundary effects such as pollution and flooding and that these effects have the potential to endanger large number of citizens.

Since the early 1990s, an environmental law framework has been established in China, with well over 100 environmental laws and hundreds of regulations, which have also provided guidance for governments to make fair decisions regarding water resource management. Apart from the laws that specifically apply to water, some general environmental legislation also applies to water management. As prescribed by a series of legal documents, including the Environmental Impact Assessment Act (EIA) and the newly promulgated Environmental Strategic Planning Law,[10] individual citizens must be informed and involved in environmental decision-making through public consultation. Both of these legal documents have been strictly applied to the provision of freshwater in urban areas, where pricing reform is being undertaken. Through the organization of public hearings, Chinese citizens now have increased opportunities to be formally involved in environmental governance (Zhong and Mol, 2008). Some researchers view this as positive sign that a more inclusive approach to environmental governance is being introduced as part of a broader state adherence to ecological modernization (e.g. Mol, 2009). To many researchers, however, these are typically token gestures that fail to promote an effective participatory agenda (Zhao, 2010; Xie, 2015). It is argued that these laws are effective measures that facilitate government agencies in reviewing and correcting problems, and ultimately promote equity of water use among the country's citizens.

China's legal framework has seen rapid developments regarding the scope of laws and the incorporation of public participation. In recent years, the state has also elaborated its environmental legislation, purportedly allowing even more space for citizens to participate in environmental governance through information sharing. For instance, the MEP established the Measures on Environmental Administrative Reconsideration in 2008 to curb unjust treatment of citizens by government officials regarding matters of environmental protection.[11] This law expands the legal framework of the Administrative Reconsideration Law and the Regulations for the Implementation of Administrative Reconsideration Law,[12] providing guidance for the administrative reconsideration work at MEP and its subordinate environmental protection agencies. The Measures on Environmental Administrative Reconsideration now covers a broader range of areas upon which citizens can seek judicial relief against unsatisfactory administrative decisions. In China's legal documents, new developments such as the Environmental Protection Law (2014) declare that individual citizens are 'entitled' to environmental information, and other documents such as the EIA Act and the Environmental Strategic Planning Law allow for participation in public consultations. Yet this legislation is criticized for not opening up full and genuine public access to official decision-making, and for neglecting to incorporate a clear set of environmental rights for the public (Xie, 2014, 2015).

## Conclusion

This chapter started by reviewing China's water policy and principles. By analysing national water uses, it pointed out that the country has enjoyed a secured level of water use ratio on most of its rivers, including those shared with its riparian neighbours. As the Chinese authorities have adopted a strong technocratic approach, its overall water security has strongly influenced how the country appropriates water that it shares with neighbouring states.

However, China's evolution in water management has been complex. Reforms have been undertaken with regard to the fundamental principles of water use. As prescribed in China's water laws and policy documents, ambiguity exists in how political authorities fairly undertake administrative activities and arrangements, which have shown limitation in providing guidance to allocate and access water resources fairly. China's complicated institutional arrangements are also featured by a centralized governance structure. Its central–local government relations have a strong influence on the planning and management of water basins, including the level of authority for policy implementation and involvement of various power centres in decision-making.

In recent years, with greater demand for efficiency in its water governance, China has experienced dramatic changes in the planning and management of its water resources. Not only have new legislative arrangements and policy targets been implemented, but the central government has also begun to review its fundamental approach to accessing and exploiting water resources. Similarly, with the development of strengthened policy targets, central authorities also aim to establish effective transboundary basin management across the country. Nevertheless, domestically, the system of water management has shown limitations in the planning and management of transboundary water basins. For instance, there has been limited incorporation of basin-scale stakeholders and their interests, giving rise to increasing conflicts. It is no surprise that similar issues appear at a transnational scale. With little institutional reform undertaken, the future effects of such policy changes therefore remain to be seen.

## Notes

1 For instance, industry's consumption rate and domestic water consumption rate is approximately 5 per cent for each use, while agriculture reaches 70 per cent.
2 Water Law of the People's Republic of China was adopted in 1988. Revisions were made in 2002.
3 According to this law, no. 32, a water permit is usually valid for five years. Renewal is required before the permit expires.
4 For instance, Agreement 21 in the Water Law states that it is important to pay attention to water ecology, particularly in desert areas.
5 Environmental Protection Law, no. 130, art 29 and 31.
6 The Protection of the Ecology of the Ili River, promulgated in 2011. See http://law.riskmw.com/xinjiang/2012/01–13/mw113409_4.html [in Chinese] (accessed 14 March 2017).
7 The No. 1 policy document from the State Council, Promulgating the Three Red Lines policy (TRL) in 2011 www.moa.gov.cn/ztzl/2016zyyhwj/hgyhwj/201301/t20130129_3209963.htm (accessed 14 March 2017).
8 Promulgated by the State Council, this regulation's Article 5(2) indicates that provincial government is the delegated authority on how water access is determined.
9 This concept has been developed in Western Europe and North America. It refers to the cooperation of multiple actors in consensual problem-solving in institutions located at specific scales.

10 These are complemented by Environmental Protection Administrative Licensing Hearings Provisional Measures (2004) and Provisional Measures for Public Participation in Environmental Impact Assessment (2006).
11 This law replaces the Measures for Environmental Administrative Reconsideration and Response to Administrative Lawsuits that was promulgated by the State Environmental Protection Administration in December 2006.
12 This Law of the People's Republic of China on Administrative Reconsideration was adopted by the Standing Committee of the National People's Congress on 29 April 1999, effective 1 December 1999. Regulations for the Implementation of the Law of the People's Republic of China on Administrative Reconsideration are promulgated by the State Council on 29 May 2007, effective 1 August 2007.

## References


Benson, D., Jordan, A. and Smith, L. 2013. Is environmental management really more collaborative? A comparative analysis of putative 'paradigm shifts' in Europe, Australia and the United States, *Environment and Planning A*, 45(7): 1695–1712.

Cai, S. 2007. *The Ethics and Legislation of River Management* [*Heliu lunli yu heliu lifa*], Henan: The Yellow River Water Conservancy Press.

Cheng, H. and Hu, Y. 2012. Improving China's water resources management for better adaptation to climate change, *Climatic Change*, 112(2): 253–282.

Chi, C. 1963. *Key Economic Areas in Chinese History as Revealed in the Development of Public Works for Water Control*, New York: Paragon Books.

Chien, S.-S. 2010. Economic freedom and political control in post-Mao China: A perspective of upward accountability and asymmetric decentralization, *Asian Journal of Political Science*, 18(1): 69–89.

Chu, Y., Hipel, K., Fang, L. and Wang, H. 2014. Systems methodology for resolving water conflicts: the Zhanghe river water allocation dispute in China, *International Journal of Water Resources Development*, 31(1): 106–119.

Dai, L. 2015. Exploring China's approach to implementing 'eco-compensation' schemes: The lake Tai watershed as case study considered through a legal lens, *Water International*, 39(5): 755–773.

Holdaway, J. 2010. Environment and health in China: An introduction to an emerging research field, *Journal of Contemporary China*, 19(63): 1–22.

Hu, Y. and Cheng, H. 2013. Water pollution during China's industrial transition, *Environmental Development*, 8: 57–73.

Huang, X. 2004. International principles on the uses and the protection of watercourse [Lun guoji shuiyu liyong he baohu de yuanze ji dui woguo de qishi], *Science Technology and Law*, 1: 96-99.

Jia, S. 2012. How to prevent the 'Three Red Lines' turn wrong? Water Resource Management Working Group, Chinese Hydraulic Engineering Society Annual Meeting.

Jia, S. and Zhang, S. 2000. When will fresh water use in China reach the climax? *Advance in Water Science*, 11(4): 470–477.

Jia, S., Lv, A., Han, Y., Long, Q., Zhu, W. and Yan, H. 2014. *Water Resources Security Report of China*, Beijing: Science Press.

Jia, S., Sun, Y., Svensson, J. and Mukherjee, M. 2016. Comparative analysis of water rights entitlements in India and China, *Water Policy*, 18(S1): 50–67.

Jia, S., Yang, H., Zhang, S., Wang, L. and Xia, J. 2006. Industrial water use Kuznets Curve: Evidence from industrialized countries and implications for developing countries, *Journal of Water Resources Planning and Management*, 132(3). Available at: http://

ascelibrary.org/doi/abs/10.1061/%28ASCE%290733-9496%282006%29132%3A3%28183%29 (accessed 10 July 2017).

Jiang, Y. 2015. China's water security: Current status, emerging challenges and future prospects, *Environmental Science & Policy*, 54: 106–125.

Li, X., Chang, F. and Chen, J. 2011. Summary of researches on water rights and initial water rights allocation, *Journal of Yangtze River Scientific Research Institute*, 28(2): 1–9.

Liu, B. and Speed, R. 2009. Water resources management in the People's Republic of China, *International Journal of Water Resources Development*, 25(2): 193–208.

Mol, A. 2009. Urban environmental governance innovations in China. *Current Opinion in Environmental Sustainability*, 1: 96–100.

Moore, S. 2014. Hydropolitics and inter-jurisdictional relationships in China: The pursuit of localized preferences in a centralized system, *The China Quarterly*, 219: 760–780.

MWR (Ministry of Water Resources). 2014. *Annual Report*, Beijing: MWR.

MWR (Ministry of Water Resources). 2015. *Annual Report*, Beijing: MWR.

MWR (Ministry of Water Resources). 2016. *Annual Report*, Beijing: MWR.

Pohlner, H. 2015. Institutional change and the political economy of water megaprojects: China's South-North water transfer, *Global Environmental Change*, 38: 205–216.

Qian, Z. 1985. Full service and transit to create new situation of water works, *Harnessing Huai*, 4: 3–4.

Shan, P. 2016. Legislation on the acquisition of water rights and their priorities, *Tsinghua University Law Journal*, 10(1): 142–159.

Shapiro, J. 2012. *China's Environmental Challenges*, Cambridge: Polity Press.

Shen, D. 2014. Post-1980 water policy in China, *International Journal of Water Resources Development*, 30(4): 714–727.

Silveira A. and Richards, K. 2013. The link between polycentrism and adaptive capacity in river basin governance systems: Insights from the River Rhine and the Zhujiang (Pearl River) basin, *Annals of the Association of American Geographers*, 103(2): 319–329.

Spijkes, O., Li, X. and Dai L. 2014. Sustainable development in China's international and domestic water law, *The Journal of Water Law*, 24(5/6): 207–219.

Wang, C. 2008. *Competition and Interdependence in Regional Cooperative Administration: Empirical Study of Urban Areas in Yangtze River Basin [Jingzheng yu yicun zhong de quyu hezuo xingzheng: jiyu changjiangsanjiaozhou dushiquan de shizheng yanjiu]*, Shanghai: Fudan University Press.

Wang, H. and Wang, J. 2012. Sustainable utilization of China's water resources [zhongguo shuiziyuan yu kechixu fazhan], *Bulletin of Chinese Academy of Sciences*, 3: 352–358.

Wang, Q., Zhou, X. and Hu, X. 2007. Consultation mechanism framework of inter-regional water resources management in China, *Advances in Science and Technology of Water Resources*, 27(5): 80–84.

Wang, S. 1999. Realizing the transition from engineering oriented water resource development to resources-oriented water resources development to do the better paper of water sector facing 21st century, *Water Economics*, 4: 1–6, 17.

Wang, S. 2000. Water rights and water market, *Journal of China's Water Resources*, 11: 6–9 [in Chinese].

Wang, Z. and Ma, L 2014. Fiscal decentralization in China: A literature review, *Annals of Economics and Finance*, 15(2): 751–770.

Wei, T. 2011. Development of China's water culture concepts viewed from water, *Journal of China University of Mining and Technology* (Social Science), 2.

Weingast, B. 2009. Second generation fiscal federalism: The implications of fiscal incentives, *Journal of Urban Economics*, 65(3): 279–293.

Winalski, D. 2009. Cleaner water in China? The implications of the amendments to China's law on the prevention and control of water pollution, *Journal of Environmental Law and Litigation*, 24: 181–202.

Xie, J., Liebenthal, A., Warford, J.J., Dixon, J.A., Wang, M., Gao, S., Wang, S., Jiang, Y. and Ma, Z. 2009. *Addressing China's Water Scarcity: Recommendations for Selected Water Resource Management Issues*, Washington, DC: World Bank.

Xie, L. 2014. Social movements and political citizenship in China, in H.-A. van der Heijden, ed., *Handbook of Political Citizenship and Social Movements*, Cheltenham: Edward Elgar, pp. 547–571.

Xie, L. 2015. Political participation and environmental movements in China, in R.L. Bryant, ed., *International Handbook of Political Ecology*, Cheltenham: Edward Edgar, pp. 59–82.

Yan, J. Forthcoming. Assessment of China's transboundary water resources and water use, doctoral thesis, Chinese Academy of Sciences, Beijing.

Yang, X., Lu, X. and Ran, L. Sustaining China's large rivers: River development policy, impacts, institutional issues and strategies for future improvement, *Geoforum*, 69: 1–4.

Zhang, J. and Bai, J. 2004. Property right study in water resource allocation, *Journal of Xi'an University of Finance and Economics*, 17(6): 74–77.

Zhang, L. and Zhong, L. 2010. Integrating and prioritizing environmental risks in China's risk management discourse, *Journal of Contemporary China*, 19(63): 119–136.

Zhang, L., Mol, A. and He, G. 2016. Transparency and information disclosure in China's environmental governance, *Current Opinion in Environmental Sustainability*, 18: 17–24.

Zhao, Y. 2010. Public participation in China's EIA regime: Rhetoric or reality? *Journal of Environmental Law*, 22(1): 89–123.

Zhao, Z. 1987. Speech to the 13th National Congress of the Communist Party of China.

Zheng, Y. 2007. *De Facto Federalism in China: Reforms and Dynamics of Central–Local Relations*, Singapore: World Scientific.

Zhong, L. 2007. Governing urban water flows in China, PhD thesis, Wageningen University.

Zhong, L. and Mol, A. 2008. Participatory environmental governance in China: Public hearings on urban water tariff setting, *Journal of Environmental Management*, 88(4): 899–913.

# 3 China's policy over international rivers

## Perception and diplomatic practices

### Introduction

Transboundary rivers run across or along the national borders of neighbouring countries. Approximately 40 per cent of world's population lives in river and lake basins. In total, these basins account for nearly half of the Earth's land surface and 60 per cent of global freshwater flows.[1] Compared to most other countries, China has a large number of international rivers running through its territory. China shares rivers with 13 of its 14 neighbouring states (Table 3.1), the exception being Afghanistan. Most of the world's largest rivers originate in China. These rivers are located to the north-east, north-west and south-west of the country (see Figure 1.1) and most of their basins are found in steep terrain and remote areas that suffer from underdevelopment. These rivers can be recognized for the following features:

- Rivers in the north-east are primarily boundary rivers marking the borders between China and its neighbouring states. In this area, China shares ten boundary rivers and three boundary lakes, with a border reaching up to 5,000 km.
- In the north-west, many of the region's rivers have an outflow from China. The Ili River originates in Kyrgyzstan, it has a total length of 1,500 km, with 422 km within China. It flows out of China at Korgas, before running into Lake Balkhash. In Xinjiang, the Irtysh River originates in the Altai Mountains, and flows into Lake Zaisan in Kazakhstan before joining the Ob River in western Siberia. The total water runoff from outside China to Xinjiang amounts to 8.8 billion $m^3$, while the runoff from China amounts to 24 billion $m^3$, which is 36.7 per cent of the total runoff (including glacier runoff) from Xinjiang. Water flowing outwards across or along Chinese borders measures approximately 730 million $m^3$, more than 30 times that flowing back into China.
- Most rivers in the south-west originate in China. They include the Gyirong Zangbu River, the Majia Zangbu River, the Senge Zangbu/Shi Quan River, the Yarlung Zangbo River, the Lancang River on the upstream of the Mekong River, the Nu River on the upstream of the Salween River, the

Yuan River located on the upstream of the Red River, and the Dulong River flowing east to the Irrawaddy River whose tributaries run through China.

Indeed, China is in possession of rich water resources from the shared river basins. Its influence on water uses of transboundary rivers can be viewed from three aspects, which are: China's possession of water resources within its territory; water use, which refers to the extraction of water for irrigation and human consumption; and the ratio of its water use to the total river basins, which refers to the ratio between a country's water extraction and the total water resources. China has exhibited various extents of influence on different river basins (Table 3.2).

Yan (forthcoming) indicates that in ten transboundary river basins studied, China is not the country that possesses the largest proportion of the basin's water resources. More specifically, for about a half of the basins, water resources that are possessed within China's territory accounts for over a quarter of the basin's water resources. This is the case for the basins of the Mekong the Nu, the Red, the Amur, the Ili and the Irtysh Rivers. On the Indus, the Ganges, the Ob and the Yarlung Zangbo Rivers, the water resource supplied from China is only a small proportion of the basin's total water supply.

China's water use refers to the extraction of water for irrigation and human consumption. It is apparent that China's influence on the total extraction of water from most of the transboundary rivers remains limited. China's water has to be balanced between extraction as an economic resource and leaving enough water to maintain a river's ecology. On the Indus, Ganges, Yarlung Zangbo, Mekong and Ob Rivers, the total withdrawals by China amount to no more than about 10 per cent of the total withdrawal from the river basins.

Furthermore, on over half of the shared river basins, China uses less water than runs into the basin from its territory. For instance, on the Indus, Ganges and Yarlung Zangbo Rivers, the runoff from China amounts to 3.3 per cent, 3.0 per cent and 13.1 per cent in the river basin, while China's uses of water resources reaches to approximately 0.1 per cent, 0.6 per cent and 0.7 per cent of the basin's total water resource. Similarly, in the south-west, for the Mekong and Nu Rivers, the runoff from China amounts to 26.1 per cent and 39.1 per cent of these basin's total water resources, respectively, while China's uses of water resources reaches to approximately 3.7 per cent and 1.8 per cent of the basin's total water resource.

However, it is also noted that over the river basins in the north-east and north-west, China and neighbouring riparian states have shown varying degrees of exploitation. In most cases, China has shown a higher degree of water exploitation, such as on the Amur and Ili Rivers. On the Ili River, Kazakhstan has reached a slightly higher level (42.7 per cent) in the ratio of water resource use from the river basin compared to China's 36.2 per cent. On the Amur River, a stark contrast exists between China (40.5 per cent) and the other two riparian neighbours Mongolia (1.1 per cent) and Russia (0.9 per cent), which have barely developed plans for the management of this river (Yan, forthcoming).

*Table 3.1* Large rivers that China shares with its riparian neighbours

| *Location in China* | *Name of international rivers* | *Origin* | *Other riparian states (apart from China)* |
|---|---|---|---|
| North-east | Heilongjiang/Amur River | Mongolia | Russia, Mongolia |
| | Yalu River | Jilin | North Korea |
| | Tumen River | Jilin | North Korea, Russia |
| | Suifen River | Jilin | Russia |
| South-west | Irrawaddy River | Tibet | Myanmar |
| | Nu River/Salween River | Tibet | Thailand, Myanmar |
| | Lancang River | Qinghai | Thailand, Myanmar, Cambodia, Vietnam, Laos |
| | Zhu River | Tibet | Vietnam |
| | Yarlung Zangbo River/Brahmaputra River | Tibet | India, Vietnam, Bangladesh, Bhutan |
| | Bajilati River/Ganges River | Tibet | Nepal, India, Bangladesh |
| | Sengezangmu River/Indus River | Tibet | Afghanistan, Pakistan, India |
| | Yuan River/Red River | Yunnan, China (outflow) | Myanmar, Laos, Thailand, Cambodia, Vietnam |
| North-west | E'erqisi River/Irtysh River | Xinjiang | Kazakhstan, Russia, Mongolia |
| | Ili River | Kyrgyzstan | Kazakhstan, Kyrgyzstan |
| | Tarim River | Tajikistan | Kyrgyzstan, Tajikistan |

*Table 3.2* Water security of river basins

| *Basin name* | *Water resources ($10^9 m^3$)* | *Water resources of China ($10^9 m^3$)* | *Ratio of China to the total basin (%)* | *Total withdrawal ($10^9 m^3$)* | *Withdrawal of China ($10^9 m^3$)* | *Ratio of China's withdrawal to the total basin withdrawal (%)* | *Ratio of China's uses of water resources to the total basin (%)* |
|---|---|---|---|---|---|---|---|
| Indus River | 1,895.6 | 64.4 | 3.3 | 2,590.8 | 0.1 | 3.9 | 0.1 |
| Ganges River | 4,542.4 | 323.5 | 3.0 | 3,275.1 | 0.8 | 2.4 | 0.6 |
| Mekong River | 6,929.1 | 749.0 | 26.1 | 312.0 | 28.0 | 9.0 | 3.7 |
| Ob River | 3,850.0 | 101.7 | 2.6 | 234.1 | 24.1 | 10.2 | 23.7 |
| Red River | 1,231.0 | 450.0 | 36.6 | 135.1 | 28.3 | 21.0 | 6.3 |
| Nujiang/Saerwenjiang | 1,739.4 | 720.9 | 39.1 | 38.6 | 13.1 | 33.9 | 1.8 |
| Heilong (Amur) River | 3,465.0 | 1,122.3 | 32.4 | 474.3 | 454.2 | 95.8 | 40.5 |
| Yarlung Zangbo River | 9,425.4 | 3,334.3 | 13.1 | 379.8 | 24.7 | 6.5 | 0.7 |
| Ili River | 241.1 | 156.9 | 65.1 | 92.7 | 56.8 | 61.3 | 36.2 |
| Irtysh River | 377.6 | 100.8 | 26.7 | 49.9 | 15.4 | 30.9 | 15.3 |

Source: Yan (forthcoming).

## China's exercising of rights and responsibilities on environmental issues

This section reviews China's perception of its sovereign rights and their application over shared rivers. Since the United Nations' Conference on the Environment and Development, environmental foreign policy has been incorporated into China's foreign policy agenda. As in many industrialized countries, China's participation in international environmental cooperation has involved the adoption of a sovereignty principle to protect the country's national interests in economic growth (Eckersley, 2004). Yet China has shown an evolving attitude towards international society on global environmental issues, hence it adjusting its position on the application of its sovereign rights.

In 1954, China developed five principles for peaceful coexistence in its foreign policy. These principles advocate: mutual respect for each other's territorial integrity and sovereignty; mutual non-aggression; mutual non-interference in each other's internal affairs; equality and cooperation for mutual benefit; and peaceful coexistence. Essentially, the country persists with these principles to advocate autonomy in its domestic affairs and in its engagement in international environmental cooperation. In particular, China persists with exercising its sovereignty rights to access and use natural resources. China aligns these rights with the United Nations' 1972 agreement on the environment, which emphasizes that states should preserve their permanent sovereignty over natural resources (Perrez, 1996; Eckersley, 2004). In Principle 21 of the 1972 Stockholm Declaration (and reaffirmed in Principle 2 of the 1992 Rio Declaration on Environment and Development), it states that 'States, have, in accordance with the Charter of the United Nations and the principles of international law, the sovereign right to exploit their own resources'.[2] Similarly, since the 1980s, under the Li Peng government, Chinese authorities have also reiterated the country's position, stating that the sovereignty of rights to natural resources must be respected. No country can interfere with the decisions of another with regard to the use of natural resources. Similarly, China insists that environment and development should be integrated, but environmental protection should not be achieved at the expense of the economy. With these sovereign rights, the country advocates that it enjoys absolute authority in the appropriation of natural resources, and the involvement of international actors is regarded as unwelcome intervention in China's domestic affairs (Li, 2015).

In recent years, China has become more open to international society and changes have been evident in the country's exercising of the sovereignty principle (Ding, 2008). Since the 1980s, Chinas had placed the environment firmly on the international agenda. It had participated in an increasing number of major international environmental norms relating to natural resources, species and habitats; stratospheric and climate change; and air pollution, hazardous waste and toxic chemicals (see Table 3.3). Since China's material power has grown, the country has also shown a capacity and willingness to commit to voluntary regulation in its environmental governance. It is observed that China

*Table 3.3* China's engagement with United Nations environment norms

| *Areas of environmental agreement* | *Year China joined the treaty* | *Name of environmental agreement and year of signing* |
|---|---|---|
| Natural resources, species and habitats | 1980 | International Convention for the Regulation of Whaling (1946) |
| | 1980 | International Convention on Civil Liability for Oil Pollution Damage (1969) |
| | 1981 | Convention on International Trade in Endangered Species of Wild Fauna and Flora (1973) |
| | 1991 | Basel Convention on the Control of Transboundary Movements of Hazardous Wastes and their Disposal (1989) |
| | 1992 | Convention on Biological Diversity (1992) |
| | 1993 | Convention on Wetlands of International Importance (Ramsar Convention, 1971) |
| Stratospheric and climate change | 1992 | United Nations Framework Convention on Climate Change (1992) |
| | 1998 | Kyoto Protocol (1997) |
| | 2009 | The Copenhagen Accord (2009) |
| Air pollution, hazardous waste and toxic chemicals | 1991 | Montreal Protocol (1987) |

has begun to act as leader in negotiations and in regional multilateral environmental cooperation (Yang, 2014). Various pragmatic approaches have been adopted. Regionally, China has been an active participant in regional multilateral cooperation. The country has been involved in the Asia-Pacific Economic Cooperation Center for Environmental Protection, and has cooperated in North-East Asia, the East Asian Sea Action Plan, and the North-West Pacific Action Plan.

It is noted that China has displayed a positive attitude towards engaging in international cooperation (Hill, 2015), albeit in selected areas (Harris *et al.*, 2013). China has welcomed financial assistance and policy transfers on environmental management from international partners such as the Netherlands, Germany, the UK and the USA. The country has also greatly benefited by collaborating with a number of United Nations bodies;[3] either in institutional reforms on the country's policies, or by promoting environmental policy instruments and technological developments in environmental management (Mushkat, 2010). In comparison, the country has shown an insufficient commitment on issues that are deemed to be constraints to national development. On the issue of banning chlorofluorocarbons (CFCs) to prevent the depletion of atmospheric ozone levels, China has hesitated without the provision of financial assistance. Hence, at times, the sovereignty right is also referred to as a 'principled strategy' (Stalley, 2013) which facilitates China's self-interests or functions as a bargaining chip for better conditions in regime formation (Saul, 2013).[4]

Along with increasing influence in the international community, China has begun to focus more on acting as a responsible superpower. Strategically, this reflects the country's intention to project a positive image in the international community (Yu, 2006; Yeophantong, 2013). Being a superpower, China's behaviours are closely scrutinized for their impact on regional stability and prosperity. As Chinese domestic scholars and policymakers have feared, the language of 'responsibility', apart from its potential to establish a positive image for China, may well undermine the country's freedoms in international affairs (Li, 2008; Yeophantong, 2013). This is most visible in climate change negotiations. China had been allied with industrializing states to pursue common interests. However, along with its growing economic power, China's identity as a developing state has been increasingly questioned. The country's perception of responsibility as a large developing state is no longer interpreted in the same way by its partners in the global South (Liu, 2015). Similarly, different interpretations of 'responsibility' have also become evident in the exploitation of natural resources (Shambaugh, 2011). This paradox poses challenges for the country's participation in international environmental cooperation.

Nevertheless, despite the government displaying an intention to share international obligations (*guoji zeren*), the extent that it shows responsibility along its development path has also become heatedly debated. For instance, it has been observed that China has declined to take an equal share in world affairs as is expected from a fair stakeholder (Yeophantong, 2013). China faces a dilemma in acting as a responsible partner in international environmental cooperation.

## China's exercise of its sovereign rights over water

China's position towards water is rather ambiguous, being a combination of its foreign policy principles and domestic water policy concerns. Although the country's domestic water policy has identified a number of key concerns for water resource management, how these two policy areas combine has proven to be complex. This section reviews China's foreign and domestic water policies in relation to the management of transboundary rivers.

China has deliberately developed varying levels of diplomatic cooperation, including forming water data sharing agreements with downstream countries (see Table 3.4). Due to a lack of regional cooperation, institutionally, a low level of institutionalization is likely to be achieved in setting up basin-wide water management on rivers that China shares with its neighbours. The more shared basins China has, the greater the number of agreements with the relevant country (He *et al.*, 2014).

Judging by the exercising of its sovereign rights, China has shown an ambiguous attitude towards the management of international rivers. Its application of sovereign rights reflects its intention to maximize profits from the use of international

*Table 3.4* China's bilateral mechanisms on transboundary water cooperation

| *Country of interaction* | *Number of shared rivers* | *Number of agreements* | *Economic use (transport, hydropower generation)* | *Cooperation disaster prevention* |
|---|---|---|---|---|
| Russia | 8 | 13 treaties | Yes | Yes |
| North Korea | 4 | 11 treaties | Yes | Yes |
| Kazakhstan | 2 | 9 treaties | Yes | Yes |
| Mongolia | 3 | 4 treaties | Yes | Yes |
| Myanmar | 3 | 4 treaties | Yes | Yes |
| Laos | 2 | 3 treaties | Yes | Yes |
| Vietnam | 3 | 2 treaties | Yes | Yes |
| India | 2 | 3 MOUs | No | Yes |
| Bangladesh | 1 | 1 MOU | No | Yes |
| Afghanistan | 3 | None | – | – |
| Bhutan | 2 | None | – | – |
| Kyrgyzstan | 1 | None | – | – |
| Nepal | 1 | None | – | – |
| Pakistan | 1 | None | – | – |

rivers. From China's participation in water-related environmental agreements, the country has shown a somewhat more cooperative attitude than that shown on the protection of natural resources. China has participated in a number of water-related environmental agreements, the main ones being the Convention on Wetlands of International Importance (Ramsar Convention) and the Convention on Biological Diversity. These agreements do not have immediate impact on the management of rivers, but are relevant in that they provide guidance on the wise use of wetland resources and the protection of biodiversity respectively.

The country has applied its domestic water and environmental laws on occasions when disagreements have arisen over water sharing (Huang and Wang, 2009; Kong, 2010). As mentioned in Chapter 2, China applies its domestic perceptions of water rights in the case of international water sharing. The principle of 'first serve, first use' is adopted, although no official legal document confirms this rule. This principle justifies China's taking of a head start in the use of water resources that run within its borders. A realism perspective is adopted on water sharing. It is argued that poor and developing countries are likely to temporarily fail to insist on their rights to water resources, although this does not mean that they have lost their entitlement to the resources. And it is also argued that those capable of appropriating water are to voluntarily respect other nations' interests and avoid generating negative impacts (Hao, 2011). Hence, in the appropriation of water resources, a proactive position is reflected in debates among experts, academics and policymakers. The Chinese central authorities thus maintain that it is rational for water projects such as dam constructions, to be launched, since the country has developed its financial and technological capacity to better manage its water resources.

At the same time, China realizes that its unilateral moves towards shared water resources are likely to be questioned. With an ambiguous attitude, it has not fully embraced mainstream water norms in relation to international river sharing (such as those outlined in the UNWC and the UN Convention on Law of Non-Navigational Use of International Watercourses 1997). It has instead shown a rather reserved position towards international water norms, but in many aspects the country has adopted the key principles in these norms (Wouters and Chen, 2015).

To defend its national interests in exploiting economic values from water resources (Saul, 2011), China has questioned current international laws on the governance of transboundary water resources such as the UNWC. It has emphasized that its policies and plans over the development of transboundary resources are legitimate and rational. In particular, China emphasizes that as an upstream country, its right to appropriate water resources is not protected. According to Article 20 of the UNWC, 'Watercourse States shall, individually and, where appropriate, jointly, protect and preserve the ecosystems of international watercourses'. On the grounds that it impedes the country's sovereign territorial rule, China disagrees with the convention on establishing collaborative water management as the suggested solution intended to prevent harm. It expresses concern in how upstream countries are disadvantaged in the current international legal framework, which favours the downstream country. Downstream countries are empowered to impinge on upstream countries by demanding transparent information disclosure (Salman, 2010). Inequity may occur when upstream countries are forced to disclose information that may undermine the planning and management of their own water resources (Wegerich and Olsson, 2010). In such a situation, China believes its autonomy in exercising its sovereign rights would be violated (Ma, 2007; He *et al.*, 2014; Yang and Zhao, 2014). From this perspective, domestic advocates are concerned that the global water norms are not useful for the purposes of resolving water disputes in practice (Hu and Zhang, 2014).

### No-harm principle

Different countries have developed differing water governance systems. China's concerns about unwelcome intervention of its water management are reflected in its water management regulations. Much emphasis is placed on how the country manages international rivers and that the government strictly complies with its domestic water laws. There is increased attention given to aligning these domestic laws with global water norms in order to avoid negatively impacting on other countries (McIntyre, 2014; He, 2015; Huang, 2015).

China stresses that sensible measures have been taken in its appropriation of water from transboundary rivers and that it respects the interests of lower riparian states. China's management of its shared rivers is primarily informed by the country's domestic water laws (see Chapter 2 for details). For instance, it considers the underlying no-harm principle when practising scientific decision-making. The country stresses that its decisions are rational and optimal because they are informed by scientific evidence. To demonstrate that the nation respects other countries, China's Ministry of Foreign Affairs stipulates that new transboundary river projects must undergo a thorough scientific assessment, with consideration of the interests of both downstream and upstream riparian countries (He *et al.*, 2014). The ministry also specifies such projects must comply with the government's transboundary environmental impact assessments (TEIAs), which is viewed as an important principle in the UNWC (He *et al.*, 2014). For instance, the recent Three Red Lines (TRL) policy has also incorporated the principle that evidence-based scientific decision-making must underpin water resource management (GWP, 2015), and that TEIAs be adopted in governmental planning and management of water resources (GWP, 2015). However, as has been shown in domestic EIAs, such measures can serve to provide additional means of legitimizing the central authorities' decisions. In practice, the EIA is not well implemented and this has raised claims of procedural injustice (Xie, 2011). Similarly, without establishing regional institutions, the country has shown few attempts to implement TEIAs in its domestic water policy.

Similar to its concerns regarding domestic water management, China focuses on the prevention of physical threats posed to humans, including through flooding, drought and water pollution. China advocates that it prioritizes human security, and the prevention of disastrous events is regarded as a primary obligation to fulfil in its cooperative activities (He, 2008). Its foreign policy agenda incorporates commitments to protect the populations of its neighbouring riparian states and peacekeeping to avoid security threats from water disasters. For instance, in an early dispute on the Mekong River, intense challenges arose not only from the Mekong River Commission and its members, but also international society.[5] In 2010, many regions in south-west China and in other countries in the Mekong basin experienced droughts. China has managed to allocate water resources to countries that experience serious droughts. It is noted that disaster prevention accounts for most of the institutional cooperation that China has developed with its neighbouring riparian states (see Table 3.4).

While China has shown a willingness to commit to global water norms, including through its no-harm and collaboration principles, in practice its interpretation of these norms is highly focused on domestic water concerns and associated policies. A high degree of flexibility is often displayed in how China implements such global water norms, with economic interest being a key consideration. Although Chinese authorities are developing plans for the sustainable use of shared waters, only a small number incorporate content on the protection of water quality and ecology or aquatic biodiversity conservation (He and Feng, 2006; Huang and Wang, 2009). These considerations are not positioned as primary concerns. For instance, as in domestic water management, biodiversity conservation has only been incorporated since the mid-2010s, and it appears that this is also the case with international river management.

## The implementation of water diplomacy

Aiming to fulfil its interests in international water resources, China is aware of tensions that escalate over its unilateral actions on the planning and use of transboundary rivers. With a series of cooperative initiatives, the country is keen to maintain friendly foreign relations with its neighbouring states. However, as mentioned above, China's position over the sharing of transboundary water resources prioritizes its national interests. On the one hand, it insists that it is entitled through its sovereign rights to appropriate transboundary water resources. But on the other hand, it disregards the relevant norms such as prior consultation and information sharing that are deemed necessary for such appropriations. The country's ambiguous attitude towards global water norms has resulted in inconsistent diplomatic practices.

In a pragmatic stance, China's water diplomacy emphasizes the resolution of conflicts, particularly those that are likely to threaten regional stability. It has displayed inconsistent positions (Huang and Wang, 2009) and placed great emphasis on the water issues that have the potential to impact on foreign relations. Border rivers, in general, receive more diplomatic attention than transboundary rivers (Wouters and Chen, 2011). Rivers that trace international borders are likely to be a cause of concern in foreign relations, such as between China and Russia (Vinogradov and Wouters, 2013). Often, these concerns result from the historical evolution of interstate interactions over territorial issues. Therefore, China's emphasis is placed on resolving water management issues (and other sensitive matters) through negotiation of border treaties. In many of the water sharing agreements formed, China shows only a partial commitment to sharing water information with countries with which it shares rivers (Wouters and Chen, 2011). Specific treaties dealing with this kind of cooperation have only developed in response to increasing disputes. It is argued that China is likely to achieve peaceful outcomes from such cooperative diplomatic activity, rather than using violent means to tackle disagreements and tensions over water sharing.

The Chinese central government has also been criticized for the limited responses it has offered to the increasing disputes. They have responded ineffectively to accusations, relying solely on media conferences to indicate China's intention to solve water sharing issues through consultation.[6] For instance, when responding to the criticism of China's behaviour over the Mekong River, Premier Wen Jiabao remarked that regional states should 'keep to the path of pursuing common development and prosperity' and that they should do so by treating each other with 'sincerity' and through 'mutual trust'.[7] Similarly, Chinese authorities are left to repeatedly emphasize that they are rational in their scientific decision-making. However, with limited communication on its water projects such as dam constructions, its sporadic information disclosure was hardly considered convincing. On several occasions, China has issued only verbal denials to questions on its upstream policy decisions, with little communication about its non-disclosed water plans. The Chinese authorities' credibility was exposed to questioning by those accusing it of arbitrary water use towards downstream countries.[8]

### *Mechanisms and forms of diplomacy*

Concerns exist about China joining global water norms, because the country has developed only limited forms of cooperation over transboundary rivers. Bilateral agreements are the main mechanism adopted by China to settle water conflicts (see Table 3.4). As previously mentioned, apart from border treaties that incorporate prescriptions on shared water management, such institutional arrangements usually see a low level of formalization as a result of China's rather unsystematic diplomatic practices (He, 2015).

National government authorities are likely to be the main actors in establishing forms of administrative regulation. To ensure implementation, joint bodies or formal arrangements to implement agreed resolutions are established (see Table 3.4). This is seen in water agreements with Mongolia, Russia, Kazakhstan and India. These joint committees have been well implemented. Informal diplomatic initiatives such as consultation represent another approach to settling disputes in the co-management of shared water resources (Chen *et al.*, 2013; Han, 2015). It is also noted that China has developed informal and project-specific consultation with countries with which it has developed trusting foreign relations. Often, such mechanisms do not incorporate core issues in water resource management as incorporated in bilateral water treaties, including water allocation, benefit sharing and integrated water resource management (see Feng *et al.*, 2015). Although some argue that this flexible approach to diplomatic cooperation suits China best, it is not always appreciated. In fact, informal communication has been seen in only a few cases when China and its partnering states have developed relatively substantial levels of trust.

Since the State Council set up major reforms with the promulgation of the Three Red Lines policy (TRL) in 2011, the TRL has become a centralized policy. Along with other water laws, it provides policy guidance to Chinese authorities for the planning and management of international rivers. Most importantly, it clarifies the main bureaucratic units which are to regulate international water resources. According to the TRL, the State Council has absolute discretion over the planning and management of international rivers, including monitoring of water quality and the allocation of rights to access and use water resources (GWP, 2015).

Despite implementation of the TRL, there remains a lack of coordination in the management of water resources in China. As with domestic water management, the MEP and the MWR are the two principal government units responsible for water management. Apart from these government units which provide professional advice on water sharing, a new unit has been set up under the Ministry of Foreign Affairs (MFA). A new institution, the Department of Boundary and Ocean Affairs (DBOA), was established under the MFA in 2009. However, as mentioned in Chapter 2, various authorities and multiple government units have discretion to exploit and regulate international water resources. Each of the central government units focuses on its dedicated area without demonstrating collective or collaborative action. The DBOA was not institutionally designed to facilitate coordinated actions across the relevant ministries and local authorities. Hence, fragmented authorities are recognized in the managing and planning of water resources in China, which results in poorly implemented integrated water resource management (Feng *et al.*, 2015).

### *Economic measures in conflict resolution*

Economic compensation is another form of cooperation that China is likely to commit to. Similar to China's domestic water management, market approaches have been incorporated in tackling disputes over water sharing. Eco-compensation is one of the most influential new initiatives. It aims to provide financial and economic incentives for integrated river basin management. This policy tool aims to promote sustainable development by focusing on ecological conservation. The term 'eco-compensation' generally refers to mechanisms that provide fiscal transfers for conservation and environmental management. Eco-compensation mechanisms compensate for ecological service provision or environmental damage resulting from resource use. This policy tool provides direct payments to suppliers of ecological services. Suppliers constitute government unit (i.e. national, provincial or local) authorities, and grassroots and community-level individuals and households. The public sector constitutes a main source of finance for eco-compensation. Specific sources of revenue are allocated to cover eco-compensation schemes. Related systems of natural resource environmental taxes and fees and private actors are other sources of income for these schemes. Financial transfers can also be made between regional or administrative levels of government units (Zhang *et al.*, 2010).

China is experimenting with various approaches to reduce negative impacts on the ecology of international rivers. Early policy experiments were launched to protect watershed ecological services. Since then, water-related ecosystem functions have become a key area for receiving financial compensation. The protection of water sources and the conservation of water and soil have become two foci in these policy programmes (Zhang and Bennett, 2011). Moreover, it is suggested that the country has adopted eco-compensation payments as an effective approach to promote equity in its access and management of shared water resources (McIntyre, 2014). In 2014, a new policy tool called the Water Resource Allocation scheme was adopted with the aim of enhancing regional water management. Trading water usage and wastewater discharge permits is another area of experimentation for providing financial compensation. China's historical water pricing system encourages overuse and wastage of precious freshwater supplies. The Water Resource Allocation scheme promotes a trading system that allows users with surplus water to trade with users that are experiencing scarcity or lack of water. This policy instrument is therefore designed to facilitate central government documentation, such as the TRL guidance, by allowing the government to issue an integrated quality and quantity control over water users. At the same time, wastewater treatment objectives are included as objectives for transboundary monitoring sections. By imposing monetary penalties on districts and counties that fail to protect water resources, the government is hoping to gain better control over total water use and total wastewater discharged within transboundary rivers (GWP, 2015). The eco-compensation approach can be adapted to a country's level of socio-economic development. Hence, this market instrument is seen as a reasonably established compensation system and an ideal policy tool to efficiently meet water resource demands (Jia *et al.*, 2016).

### *Information sharing and exchange in water management*

Information provision and data exchange constitutes a main form of China's cooperation with its riparian neighbours. China insists that it maintains a dialogue with relevant parties regarding the management of shared rivers, particularly when potential consequences may arise. In water management, hydrological data, water-level monitoring, and flood forecasting and mitigation are important aspects of the dialogue. This gesture signals relatively solid diplomatic cooperation between China and its riparian neighbours. To date, China has developed information sharing with seven neighbouring countries (see Table 3.5).

As already mentioned, China's cooperation places a strong emphasis on peacekeeping and the prevention of ecological disasters such as flooding, drought or emergent water pollution. This is referred to as emergency response cooperation.[9] For instance, with India, China has developed progressive diplomatic initiatives on issues of 'natural disasters'[10] and 'emergency management'. As seen in China's cooperation with the Mekong River Commission, the disclosed water data was intended to help with disaster preparedness including

*Table 3.5* Hydrological data sharing and implementation institutions established between China and selected riparian states

| *Country* | *Commission* | *Information sharing* |
|---|---|---|
| Bangladesh | None | Provision of general hydrological data and hydrological data in flood season for Yarlung Zangbo River |
| India | Joint steering commission is established | Provision of hydrological data in flood season of Yarlung Zangbo River and Langen Zangbu River |
| Kazakhstan | Joint steering commission and experts groups are established | Provision of general discharge data and data when warning level has been exceeded |
| Mongolia | Joint steering commission is established | Share hydrological data when warning level has been exceeded |
| North Korea | None | Provide hydrological data on the Yalu and Tumen Rivers |
| Russia | Joint steering commission and experts groups are established | Share hydrological data for the Amur River in flood season |
| Vietnam | None | Exchange of hydrological data |

the flash flood guidance system and flood forecasts for the Mekong River to support navigation, drought forecasting and water resources management.[11] China is unwilling to provide access to key information such as data collected from its hydrologic stations (see Gao, 2015). It has not yet developed practices to provide prior notice or consultation on planned information-sharing measures in the majority of its agreements with its neighbouring riparian states (Wouters and Chen, 2011; He, 2015).

China's sharing of such information is also constrained by domestic regulations. According to ministerial regulation,[12] information regarding international rivers is not to be made publicly transparent. In particular, when disastrous events have occurred, the lack of an established risk management system means the Chinese authorities have not managed to provide transparent information to the public (He, 2008; Zhang and Zhong, 2010). Therefore, China's pragmatic attitude towards resolving tensions has also raised concerns among domestic policymakers and elites. Domestic scholars and experts fear that China's unilateral behaviours may well leave its riparian neighbours uninformed, resulting in much distrust towards the country (Ma, 2010; Feng *et al.*, 2015).

## The incorporation of non-state actors

Existing literature identifies private economic actors as another major group that is active in public participation. This is mainly discussed by applying ecological modernization theory, based on the experiences in Western Europe, to China (Mol, 2006). In China, along with its economic growth, private corporations represent an important actor that impacts on the country's planning and management of international water resources (Zha, 2015). Non-state actors, including private economic actors, have become active in promoting ecological rationality by voluntarily internalizing environmental concerns or by forming environmental movements. However, it should be noted that the nature of the political system plays a significant role in directing and determining the outcomes of public participation in China's environmental management (Tang *et al.*, 2008; Gilley, 2012).

### *Corporate involvement in water governance*

To continue boosting the country's economic development, China has developed policies for the governance of water. In particular, a diverse water sector investment system was established to encourage local businesses to invest in water resource projects. However, Chinese authorities are faced with the challenge of how best to regulate the country's private corporations in overseas investments, with growing pressure from local protesters and international observers. Private corporations have been criticized for neglecting environmental and social impacts in their development plans.

There are two major types of venture corporations involved in water resource projects (GEI, 2015). One is listed companies, which are dominated by state-owned enterprises, and the other is limited liability firms and companies, which account for approximately half of China's total overseas investment. The latter companies pose challenges for regulation. Unlike listed companies, the limited liability firms and companies are not obliged to make certain information public, nor do they have the same incentive as state-owned enterprises to present a good image of China to the world. The challenges mainly relate to financial measures such as auditing and evaluation.

As more legal measures have been developed, China's overseas investment has become better regulated and in some cases shown improvement.[13] With domestic and international pressure, private corporations have increasingly complied with water legislation and domestic laws to facilitate equity and fairness in the implementation of water policies (Zha, 2015). The corporations are now incentivized to develop Corporate Social Responsibilities (see Box 3.1 and Table 3.6). More specifically, Chinese corporations have shown greater awareness of their host countries' environmental laws, such as requirements for environmental impact assessments, and pollutant discharge standards and emergency management procedures. State-owned corporations, in particular, have been more pressured to make changes. Sino Hydro, for instance, has implemented changes in dam construction projects to reduce their environmental impact. Nevertheless, it is noted that the enforcement of host countries' environmental laws and regulations can greatly impact on the performance of Chinese firms (Hensengerth, 2013). In situations where such regulations are absent, most corporations resort to Chinese laws as a guide. The standards applied on China's overseas projects are consistent with the standards set within China for similar projects (Woods, 2015).

### Box 3.1 The regulation of overseas investment

Chinese overseas investment is collectively regulated by the following government authorities: the State-owned Assets Supervision and Administration Commission of the State Council, the Ministry of Commerce and the National Development and Reform Commission (NDRC); and by financial institutes such as the China Exim Bank and the China Development Bank. Apart from these government units that have direct authority over Chinese private corporations, the MFA and the MEP also have discretion to provide guidance and regulation of corporations' environmental and social impacts (see Table 3.7).

Additionally, a registration and approval system has been established to supervise large-scale financial investments in natural resources, such as hydropower projects. However, some registration and approval overlap occurs between the existing political authorities. According to a new policy document promulgated by the State Council, the NDRC is the main government unit to approve overseas investment in sensitive countries and regions.[14] NDRC also has the discretion to establish departmental measures to regulate the performance of private corporations. If the investment project is not within sensitive areas, however, it has to be approved by the Ministry of Commerce. The ministry also conducts annual audits and evaluations to regulate overseas investment. However, the distinction is not always clearly made.

*Table 3.6* Legal framework supervising private corporations' overseas projects

| | *Name of regulation* | *Main focus* |
|---|---|---|
| Ministry of Foreign Affairs; Ministry of Commerce and SACSAC | Notice on Further Regulating the Foreign Investment and Cooperation of Chinese Enterprises (2008) | Requires that corporations should fulfil relevant social responsibilities, and those that have generated negative impacts would be punished financially and administratively |
| MEP and MOFCOM, Collaborative regulation | Guidelines on Environmental Protection for Overseas Investment and Cooperation (2013) | Disclosure of information, compliance with host countries' laws; and fulfilment of responsibilities such as environmental impact assessments, pollutant discharge standards and emergency management |
| MOFCOM | Measures for Overseas Investment Management (2014) | Requires that all Chinese investors observe the laws and local customs of the host states |

*Table 3.7* Authorities regulating overseas investments

| *Main authorities* | *Responsibilities* |
|---|---|
| The Ministry of Commerce (MOFCOM) | Developing management strategies and specific policies on Chinese enterprises' overseas investment, and approving and monitoring companies that are to be engaged in overseas investment |
| China Exim Bank and China Development Bank (CDB) | Providing overseas investment loans and financial assistance to develop market |
| State-owned Assets Supervision and Administration Commission of the State Council (SACSAC) | Regulating and supervising overseas state-owned assets of major state-owned enterprises |
| National Development and Reform Commission (NDRC) | Approving and document majoring projects; initiating administrative measures to regulate the use of foreign exchange in overseas investment |

### Environmental activism and environmental movement

Domestically, the Chinese government has expressed concerns about incorporating civil society actors into decision-making processes, including for issues surrounding water sharing. The Chinese central government has purposefully developed a depoliticized strategy to involve individuals and organizations in decision-making processes (Li *et al.*, 2012; Xie, 2015). Simultaneously, issues are taken out of the political sphere and presented as technical or scientific problems that require modes of handling quite different from political deliberations and contestation. This strategy seems to neutralize potential veto points in the policy implementation phase. By means of assimilation and co-optation into governmental interests, this strategy is targeted at enhancing the quality of policy implementation. The government has, however, shown ambiguous attitudes towards non-state actors' involvement in the policy process.

With regard to the organization of activism, in China and among its riparian neighbour states, changing state–society relationships are affecting the governance of various river basins, further threatening the management of transboundary river basins (Kattelus *et al.*, 2015). In China alone, diverse forms of activism exist including both legal and illegal environmental non-government organizations (NGOs), social movements, court cases and 'spontaneous' protests. Environmental activism has developed a strong focus on the conservation of rivers. In relation to tactics, transnational networks have been developed between Chinese NGOs and those in other riparian states to exert stronger pressure on the Chinese government (Morton, 2005; Xie, 2009; Scheumann and Hensengerth, 2014).

Several main streams of environmentalism have been identified as opposing the Chinese government's exploitation of rivers for hydroelectricity generation. Early campaigns focused merely on the negative ecological effects that could result from such projects. Gradually, activists began to raise public concerns about the equity and validity of an official environmental decision-making system that has rarely listened to disadvantaged groups (Xie, 2015). These concerns have been echoed by local communities in neighbouring riparian states. We are reminded that Chinese environmental activism is deeply rooted in the country's political system, which has played a significant role in shaping the environmental movement's identity and development (see also Xie and van der Heijden, 2010). China's political system has provided limited opportunities for anti-dam movements. With greater concerns for national stability and security, the Chinese Communist Party fears that political participation (both domestically or abroad) may have strong implications for promoting democracy or other moves towards political liberalization.

Similarly, those of China's neighbouring riparian states that have displayed deficient democracy have also shown varying attitudes towards grassroots environmental movements. As identified by a UN report, these states have not effectively involved the public, which has constrained efficient water management (UNEP, 2012). A case study of local protests in the East Himalaya region indicates that the bleak nature of state–civil society relations is caused by an absence of legitimate channels of articulation for local stakeholders (Huber and Joshi, 2015). Such situations have led to violence (Huber and Joshi, 2015). Additionally, the development of civil society also varies across China's neighbouring riparian states. India's experience has suggested that not all social groups have access to movement resources or are competent at articulating and practising environmental justice activism (Williams and Mawdsley, 2006). In relation to water, complicated incentives exist for the public to engage in policy processes. Communities and environmentalists may display a narrow concern which does not go beyond their local jurisdiction and the security of its water resources.[15] This explains why neighbouring states' environmental movements have achieved limited success in their opposition to China's damming activities (Zawahri and Hensengerth, 2012).

## China's position on international norms: a summary

In response to demands for improved domestic water management, China has committed to extending its domestic water policy over international rivers. The country has developed diplomatic cooperation that aims to mitigate distrust and tensions in foreign relations. There have been positive developments in China's water management practices, including the use of various policy tools to resolve disputes over water sharing. On a number of occasions, China has shown a commitment to work with neighbouring riparian states to conduct collaborative water management. Numerous policies have been developed to address issues arising from the planning and management of shared water resources (Feng *et al.*, 2015). More flexibility has also been displayed in the country's attempts to adapt to the interests of neighbouring riparian states. To relieve strains over its foreign relations, China has implemented policy changes that signal a move away from the unilateral stance as found in previous years (Aviram *et al.*, 2014). China has also shown a commitment to complementing these policies with political, social and economic inputs. However, it is noticeable that its stances are rather ambiguous and the country has not yet formed systematic principles to respond to issues of concern being raised by its neighbours.

China's endeavours to develop effective diplomatic initiatives with neighbouring riparian states have, however, proved to be increasingly troublesome. The country lacks a systematic approach for tackling water sharing issues with neighbouring states. In its policies, China has illustrated intent to mitigate environmental degradation in the rivers it shares with other riparian states (Chen *et al.*, 2013; Lee, 2014). However, weaknesses have been shown in China's dealing with the negative effects of its appropriation of transboundary water resources. Institutionally, there are limited formal arrangements in place to reduce ecological impacts. As shown in its domestic water management, China lacks robust systems and administrative apparatus to implement such policy concerns. Additionally, the country has not developed a holistic water management policy that aims to coordinate its neighbouring riparian states' interests at the river basin scale (Feng *et al.*, 2015). Confronted with the task of dealing with these riparian states' varying interests, China has shown only a conservative stance that emphasizes the peaceful settlement of disputes between concerned parties or countries. From this perspective, China's international water policy is viewed as temporary.

In the appropriation of its shared water resources, China finds itself increasingly pressured when pursuing its national interests. On the one hand, China's unilateral actions are no longer sufficient to meet its domestic water pressures as the country faces increasing uncertainties, particularly surrounding the control of disasters and water quality in rivers in the south-west (Feng *et al.*, 2015). On the other hand, water-related and non-water-related issues have the potential to strain China's foreign relations, risking regional stability and posing a threat to China's economic development (Huang and Wang, 2009). Again, we are reminded that industrializing countries are likely to experience more difficulties than developed societies.

Furthermore, China has encountered a dilemma in striking a balance between its rights as an upstream developing state while fulfilling its responsibilities towards downstream riparian states that are troubled by China's domestic water management issues. The country has developed a reserved stance over its shared rivers. Such a stance aims to pursue the country's national interests with the least degree of water diplomacy. It apparently intends to demonstrate the fulfilment of its obligations towards other riparian states, while minimizing collaboration with these states. However, this strategy has become increasingly challenged. China is increasingly pressured to take a collaborative stance focused on basin-wide sustainable development. It remains unclear whether the country has placed the goal of sustainable river basin management in its foreign policy agenda that directs its development of water diplomacy.

## Notes

1 UN Water. Available at: www.unwater.org (accessed 8 February 2017).
2 See full text of the Agenda 21, action plan agreed on United Nations Conference on the Human Environment, https://sustainabledevelopment.un.org/content/documents/Agenda21.pdf (accessed 20 February 2017).
3 China has worked with United Nations bodies such as the Commission for Sustainable Development, the Global Environmental Facility and the UN Environmental Programme (UNEP). Since 2013 it has been working with the World Bank to phase out ozone-depleting substances.
4 According to Economy (1999). See H. Yu, 'Green challenges in China–US relations', *chinadialogue*, 15 September 2006. Available at: www.chinadialogue.net/article/show/single/en/378-Green-challenges-in-China-US-relations (accessed 19 July 2017).
5 'UN says China dams threaten water supplies to Mekong delta farmers', *Probe International*, 28 May 2009. Available at: https://journal.probeinternational.org/2009/05/28/un-says-china-dams-threaten-water-supplies-mekong-delta-farmers/ (accessed 5 February 2017).
6 Personal interview with expert/activist Yong Yang in 2014.

7 'Build a bond of cooperation and a common homeland', Chinese premier Wen Jiabao's speech at the Third GMS Summit, Vientiane, Laos, 31 March 2008.
8 Personal interview with expert/activist Yong Yang in 2014.
9 Agreement on the Protection and Utilization of Border Waters Between China and Mongolia (1994); Cooperation Agreement on the Utilization and Protection of Transboundary Waters Between China and Kazakhstan (2001); Agreement on Reasonable Utilization and Protection of Transboundary Waters Between China and Russia (2008).
10 The Border Defence Cooperation Agreement (BDCA) between China and India was signed on 23 October 2013.
11 China has agreed to share more hydrological data with the Mekong River Commission by extending the hydrological data provision by 30 days, running from 1 June to 31 October, every year, as well as increasing the frequency of the data sharing to twice a day.
12 Regulations on State Confidentiality in Water Resources Management and its Specific Scope and Security Classification [*shuili gongzuo zhong guojia mimi jiqi juti fanwei de guiding*], 29 December 2000, parts 3 and 4.
13 'China: Not the rogue dam builder we feared it would be?' Peter Bosshard, *Internationalrivers.org*, 31 March 2010. Available at: www.internationalrivers.org/blogs/227/china-not-the-rogue-dam-builder-we-feared-it-would-be (accessed 19 July 2017).
14 The 18 November 2014 edition of a Catalogue of Investment Projects Subject to Government Approval. Sensitive regions refer to those that have experienced disagreement or dispute with China.
15 Interview with Bharat Seth, International Rivers, India, 10 April 2014.

## References

Aviram, R., Katz, D. and Shmueli, D. 2014. Desalination as a game-changer in transboundary hydro-politics, *Water Policy*, 16(4): 609–624.

Chen, H., Rieu-Clarke, A. and Wouters, P. 2013. Exploring China's transboundary water treaty practice through the prison of the UN watercourse convention, *Water International*, 38(2): 217–230.

Ding, J. 2008. Achievements and issues in China's environmental diplomacy [*Zhongguo huanjing waijiao de chengjiu yu wenti*], *Gansu Social Sciences*, 5 [in Chinese].

Eckersley, R. 2004. *The Green State: Rethinking Democracy and Sovereignty*, Cambridge, MA: MIT Press.

Economy, E. 1999. China's international environmental diplomacy, in S.S. Kim. ed., *China and the World*, 4th edition, Boulder, CO: Westview.

Feng, Y., He, D. and Wang, W. 2015. Identifying China's transboundary water risks and vulnerabilities: A multidisciplinary analysis using hydrological data and legal/institutional settings, *Water International*, 40(2): 328–341.

Gao, Q. 2015. *A Procedural Framework for Transboundary Water Management in the Mekong River Basin: Shared Mekong for a Common Future*, Leiden: Brill-Nijhoff.

GEI (Global Environmental Institute). 2015. *Understanding China's Overseas Investments Governance and Analysis of Environmental and Social Policies*, Beijing: GEI.

Gilley, B. 2013. *The Right to Rule: How States Win and Lose Legitimacy*, New York: Columbia University Press.

GWP (Global Water Partnership). 2015. *China's Water Resources Management Challenge: The 'Three Red Lines'*, technical focus paper. Available at: www.gwp.org/globalassets/global/toolbox/publications/technical-focus-papers/tfpchina_2015.pdf (accessed 10 July 2017).

Han, H. 2015. Approaches to investment in Chinese transboundary waters, *Water International*, 40(1): 71–86.

Hao, S. 2011. Law discussion on the ten major relationships that should be carefully dealt with to harmonious development of cross-border water resource, *Journal of Natural Resources*, 26(1): 166–176.

Harris, P.G., Chow, A.S.Y. and Karlsson, R. 2013. China and climate justice: Moving beyond statism, *International Environmental Agreements: Politics, Law and Economics*, 13(3): 291–305.

He, D. and Feng, Y. 2006. *Guoji Heliu Kuajing Shui Ziyuan Heli Liyong yu Xietiao Guanli [Rational Utilization and Coordinated Management of Transboundary Water Resources in International Watercourses]*, Beijing: Science Press [in Chinese].

He, D., Liu, C., Hu, J., Ji, X. and Li, Y. 2014. Progress and perspective of international river researches in China [Zhongguo guoji heliu yanjiu jinzhan yu zhanwang], *Acta Geographica Sinica*, 69(9) [in Chinese].

He, Y. 2008. The research on the legal problem of protecting transboundary water resources, *Journal of Natural Resources*, 23(6): 1099–1111.

He, Y. 2015. China's practice on the non-navigational uses of transboundary waters transforming diplomacy through rules of international law, *Water International*, 40(2): 312–327.

Hensengerth, O. 2013. Chinese hydropower companies and environmental norms in countries of the global South: The involvement of Sinohydro in Ghana's Bui Dam, *Environment, Development and Sustainability*, 15(2): 285–300.

Hill, D.P. 2015. Where hawks dwell on water and bankers build power poles: Transboundary waters, environmental security and the frontiers of neoliberalism, *Strategic Analysis*, 39(6): 729–743.

Hu, D. and Zhang, J. 2014. China ought to stay out of UNECE, *Journal of China's Social Sciences*, 20 November.

Huang, X. and Wang, K. 2009. On legislation of water resources utilization in international rivers in China [zhongguo guoji heliu shuiziyuan liyong lifa luelun], *Yangtze River Basin Resource and Environment*, 18(5) [in Chinese].

Huang, Z. 2015. Case study on the water management of the Yaluzangbu/Brahmaputra River, *Georgetown International Environmental Law Review*, 27(2): 229–250.

Huber, A. and Joshi, D. 2015. Hydropower, anti-politics and the opening of new political spaces in the Eastern Himalayas, *World Development*, 76: 13–25.

Jia, S., Sun, Y., Svensson, J. and Mukherjee, M. 2016. Comparative analysis of water rights entitlements in India and China, *Water Policy*, 18(S1): 50–67.

Kattelus, M., Kummu, M., Keskinen, M., Salmivaara, A. and Varis, O. 2015. China's southbound transboundary river basins: A case of asymmetry, *Water International*, 40(1): 113–138.

Kong, L. 2010. On China's problems in international rivers exploitation and legal thinking from different angles [Lun woguo zai guoji heliu kaifa zhong de wenti ji duobiwei falv sikao], *Law Science Magazine*, 1: 139.

Lee, S. 2014. Benefit sharing in the Mekong river basin, *Water International*, 40(1): 139–152.

Li, J. 2008. On the transformation of the international system from the perspective of the responsibility theory [Cong zeren lun toushi guoji tixi zhuanxing], *International Studies*, 1: 36–47.

Li, T.H.Y., Ng, S.T. and Skitmore, M. 2012. Public participation in infrastructure and construction projects in China: From an EIA-based to a whole-cycle process, *Habitat International*, 36(1): 47–56.

Li, X., Chang, F. and Chen, J. 2011. Summary of researches on water rights and initial water rights allocation, *Journal of Yangtze River Scientific Research Institute*, 28(2): 1–9.

Li, Z. 2015. Impact factors in China's sharing of transboundary rivers [Zhongguo kuaguojie heliu wenti yingxiang yinsu fenxi], *Quarterly Journal of International Politics*, 42: 66–92.

Liu, Y. 2015. Transboundary water cooperation on the Yarlung Zangbo/Brahmaputra: A legal analysis of riparian state practice, *Water International*, 40(2): 354–374.

Ma, B. 2010. On China's problems in international rivers exploitation and legal thinking from different angles, *Law Science Magazine*, 2010(1): 139–141.

Ma, X. 2007. Statement on the ILC's work on shard natural resources and responsibility of international organizations, *Chinese Journal of International Law*, 63(3): 769–772.

McIntyre, O. 2014. The protection of freshwater ecosystems revisited: Towards a common understanding of the 'ecosystems approach' to the protection of transboundary water resources, *Review of European, Comparative and International Environmental Law*, 23(1): 88–95.

Mol, A. 2006. Environment, modernity and transitional China: At the frontier of ecological modernization, *Development and Change*, 37(1): 29–56.

Morton, K. 2005. *International Aid and China's Environment: Taming the Yellow Dragon*, Abingdon: Routledge.

Mushkat, R. 2010. Compliance with international environmental regimes: Chinese lessons, *William & Mary Environmental Law and Policy Review*, 34(2): 493–542.

Perrez, F.X. 1996. The relationship between 'permanent sovereignty' and the obligation not to cause transboundary environmental damage, *Environmental Law*, 26(4): 1187–1212.

Salman, S.M.A. 2010. Downstream riparians can also harm upstream riparians: The concept of foreclosure future uses, *Water International*, 35(4): 350–364.

Saul, B. 2011. *China, Resources and International Law*, Sydney Law School Research Paper No. 11/82. Available at: http://ssrn.com/abstract=1954180 (accessed 11 July 2017).

Saul, B. 2013. China, natural resources, sovereignty and international law, *Asian Studies Review*, 37(2): 196–241.

Scheumann, W. and Hensengerth, O. 2014. *Evolution of Dam Policies: Current Practices and the State of the Debate*, Berlin: Springer.

Shambaugh, D. 2011. Coping with a conflicted China, *The Washington Quarterly*, 34(1): 7–27.

Stalley, P. 2013. Forum: Principled strategy – The role of equity norms in China's climate change diplomacy, *Global Environmental Politics*, 13(1): 1–8.

Tang, B., Wong, S. and Lau, M. 2008. Social impact assessment and public participation in China: A case study of land requisition in Guangzhou, *Environmental Impact Assessment Review*, 28: 57–72.

UNEP (United Nations Environment Programme). 2012. *Status Report on the Application of Integrated Approaches to Water Resources Management*, Nairobi: UNEP. Available at: www.un.org/waterforlifedecade/pdf/un_water_status_report_2012.pdf (accessed 17 August 2017).

Vinogradov, S. and Wouters, P. 2013. *Sino-Russian Transboundary Waters: A Legal Perspective on Cooperation*, Stockholm paper, Institute for Security and Development Policy. Available at: http://isdp.eu/content/uploads/images/stories/isdp-main-pdf/2013-vinogradov-wouters-sino-russian-transboundary-waters-legal-perspective.pdf (accessed 12 July 2017).

Wegerich, K. and Olsson, O. 2010. Late developers and the inequity of 'equitable utilization' and the harm of 'do not harm', *Water International*, 35(6): 707–717.

Williams, G. and Mawdsley, E. 2006. Postcolonial environmental justice: Government and governance in India, *Geoforum*, 37: 660–670.

Woods, N. 2015. Whose aid? Whose influence? China, emerging donor, and the silent revolution in development assistance, *International Affairs*, 84(6): 1205–1221.

Wouters, P. and Chen, H. 2011. China's 'soft-path' to transboundary water cooperation examined in the light of two UN global water conventions: Exploring the 'Chinese way', *Water Law*, 22(6): 229–246.

Wouters, P. and Chen, H. 2015. Editors' introduction to the China water papers: Transboundary water cooperation in Asia with a focus on China, *Water International*, 40(2): 286–296.

Xie, L. 2009. *Environmental Activism in China*, Abingdon: Routledge.

Xie, L. 2011. Environmental justice in China's urban decision-making, *Taiwan in Comparative Perspective*, 3: 160–179.

Xie, L. 2015. Political participation and environmental movements in China, in R.L. Bryant, ed., *International Handbook of Political Ecology*, Cheltenham: Edward Edgar, pp. 59–82.

Xie, L. and van der Heijden, H. 2010. Environmental movements and political opportunities: The case of China, *Social Movement Studies*, 9(1): 51–68.

Yan, J. Forthcoming. Assessment of China's transboundary water resources and water use, doctoral thesis, Chinese Academy of Sciences, Beijing.

Yang, J. ed. 2014. *China's Diplomacy: Theory and Practice*, Hackensack, NJ: World Century.

Yang, Z. and Zhao, Z. 2014. Issues in China's transboundary water resources [*Guonei kuajie shuiziyuan wenti yanjiu*], *Theory Monthly*, 2: 146–151 [in Chinese].

Yeophantong, P. 2013. Governing the world: China's evolving conceptions of responsibility, *Chinese Journal of International Politics*, 6(4): 329–364.

Yu, H. 2006. Knowledge factors and China's environmental diplomacy: United Nations Framework Convention on Climate Change and the development of China's administrative coordination regarding climate change, *Hong Kong Journal of Social Sciences*, 30: 75–96.

Zawahri, N. and Hensengerth, O. 2012. Domestic environmental activists and the governance of the Ganges and Mekong Rivers in India and China, *International Environmental Agreements: Politics, Law and Economics*, 12(3): 269–298.

Zha, D. 2015. A political ecology of hydropower development in China, in N. Matthews and K. Geheb, eds, *Hydropower Development in the Mekong Region Political, Socio-economic and Environmental Perspectives*, Abingdon: Routledge.

Zhang, L. and Zhong, L. 2010. Integrating and prioritizing environmental risks in China's risk management discourse, *Journal of Contemporary China*, 63(19): 119–136.

Zhang, Q. and Bennett, M.T. 2011. *Eco-Compensation for Watershed Services in the People's Republic of China*, Manila: Asian Development Bank.

Zhang, Q. Lin, T., Bennett, M.T. and Jin, L. 2010. *An Eco-Compensation Policy Framework for the People's Republic of China: Challenges and Opportunities*, Manila: Asian Development Bank.

# 4 Theoretical debate

## Water diplomacy

### Introduction: the nature of water diplomacy – a conceptualization

The definition of diplomacy is an evolving concept. It represents both the formation and implementation of foreign politics, international negotiations and technique of foreign policies, which is being performed by diplomats (Leonard, 2002). Diplomacy has various functions and it generally serves to manage the relations between countries, and between countries and other actors, through negotiation and coordination of interests (Barston, 1988). In a narrow sense, it refers to processes by which foreign policy is carried out (Amacker, 2011).

Diplomacy over water management refers to the rationale of countries' water-related foreign policy processes. Although nations are most likely to tackle disagreements through peaceful negotiation and seeking solutions that suit their national interests (Wolf *et al.*, 2003), however, the outcomes from such interstate interactions remain uncertain. Water diplomacy bears unique features and approaches for forming cooperative partnerships in the contextual features where it is developed. Consequently, the outcomes vary. Hence, water diplomacy is a complicated arena which involves complexity of features (Islam and Susskind, 2012).

Despite that international rivers meander through more than one nation, under international law they are managed independently by independent nation states. Though as a shared natural resource, river systems resemble a *public good* which can be exploited jointly and simultaneously by all riparian states within the river basin, unlike what is often found in situation of 'tragedy of the commons', countries are not incentivized to the same extent to form norms to govern the shared river basins (Waterbury, 1997, 2002; Ovodenko, 2014). One country's use of water may lead to predictable or unpredictable externalities which ultimately lead to rivalry. The consequences may appear in relation to water quality or water ecology. Hence, forming water diplomacy is an unsmooth process which may see cooperation and conflicts occur simultaneously in negotiating over how river basins are shared (Zeitoun and Allan, 2008). Institutional arrangements are likely to be the outcomes of water diplomacy, which are also referred to as water management institutions (WMIs). At least three separate sectoral interests can be distinguished in cooperative initiatives, which tend to focus on one or more gains from WMIs, namely, the ecological protection of river basins, economic profits from water resources and security interests from possessing water information.

Water ecology provides the fundamental basis of sustainable water management. This is because aquatic ecosystems provide for the exploitation of water. Some have argued that many countries place too high a focus on the maintenance and integrity of their ecosystems (Dietz *et al.*, 2003; Raadgever *et al.*, 2008). Successful river management in Europe has shown that collaboration based on multisectoral water management has also proved possible for shared water resources (Young, 1989; Haftendorn, 2000). Multiple economic interests (as targets) can be achieved when adaptive forms of environmental management, and advanced governance arrangements, are established (Dietz *et al.*, 2003; Raadgever *et al.*, 2008).

Economic growth remains a key issue in developing water policies. Economic interests over water resource management often represent a primary concern. Despite that countries may attempt to form cooperative institutions, such efforts do not always lead to effective resolution of problems relating to water basin management (Sadoff and Grey, 2005). More specifically, countries display varying degrees of willingness to cooperate due to their prioritization of important national interests such as economic growth (Fox and Sneddon, 2007; Molle *et al.*, 2009). For developing countries, such interests are often separated from the holistic management of water resources, such as when primary interests are shown in hydropower generation and the fulfilment of irrigation needs from water (Fox and Sneddon, 2007). These countries are likely to prioritize national interests over water basin management or transboundary interests (Hirsch and Mørck Jensen, 2006). The resultant institutional cooperation may lead to negative effects for transboundary ecology (Fox and Sneddon, 2007; Molle *et al.*, 2009).

Water resources are multifaceted and substantial political interests may be vested in countries' negotiations over water. In comparison to other issues of international relations, transboundary rivers pose greater challenges to international relations, particularly in regions where instability is seen among nation states (Lindemann, 2008). More specifically, rivalry in power relations exists because countries' asymmetrical usage may lead to negative externalities for other riparian users. Asymmetries are likely to be displayed in countries' possession of policy information over the use of shared rivers (Waterbury, 2002). This is particularly the case in upstream–downstream situations, whereupon upstream countries are viewed as possessing important information they take unilateral actions over shared waters. Furthermore, military or strategic interests may exist when countries withhold important water information from other countries. Water information has a dual nature. On the one hand, it is highly important to river basin management, as countries have strong interests in sharing information to promote establishing adaptive water governance (Dietz *et al.*, 2003; Raadgever *et al.*, 2008). On the other hand, hydrological information can be used as a 'weapon' to gain political leverage (Gleick, 1990; Timmerman and Langaas, 2005). This is often seen among upstream countries and has been described as countries 'mov[ing] ahead with projects that affect the flow or quality of the resource[s]' (Waterbury, 1997: 279). More broadly, it is referred to as a resource-capture strategy for these countries to guarantee their own access to and use of shared rivers (Waterbury, 1997). When countries bring their political interests to the negotiations, increasing disagreement over diplomacy can become a source of conflict between countries with shared water resources (Bernauer and Kalbhenn, 2010; Brochmann and Gleditsch, 2012).

Water may also be a source of interest for governments to realize domestic political interests. Literature has recorded that in the case of River Jordan conflict, both Israel and Jordan share common reasons to cheat, although their interests differ. In particular, Jordan's adoption of an environmental agreement was found to satisfy pressures stemming from domestic public opinion (Fischhendler, 2008). National governments' interests supply another distracting factor that impacts on water diplomacy. Regarding water disputes, there are suspicions that the purpose of countries' participation in international institutional arrangements is one of free-riding (Daoudy, 2007; Sosland, 2007). States, particularly those that are in the phrase of industrializing, are found to have incentives to free-ride (Fox and Sneddon, 2007; Mitchell and Hensel, 2007) or cheating in the development of WMIs (Fischhendler, 2008; Zawahri, 2008). Hence, political interests are identified as another factor that impacts on water diplomacy.

In this study, cooperation is conceptualized as state-led diplomatic cooperation on transboundary water management. It recognizes that states' incentives for water diplomacy vary widely and that the cooperative processes do not necessarily lead to positive outcomes for the management of transboundary river basins. The conceptualization differs from conventional wisdom that distinguishes formality or level of institutionalization as a symbol of cooperation. As a result, water diplomacy may be signalled by formalized or flexible mechanisms such as WMIs between two or more countries. By so doing, emphasis is placed on the nature of cooperation that may affect the outcome of diplomatic cooperation. Water diplomacy, accordingly, is recognized as involving a complexity of features that impact on whether substantial cooperation can be formed. A process-focused perspective is required that investigates the nature of cooperation exhibited by water diplomacy by linking the nation states' motivating factors, compliance with international agreements, and ultimately the outcomes of such cooperation.

Overall, the scholarship on interstate behaviours over international rivers has placed much emphasis on the conditions that facilitate the formation of international institutions (Hensel *et al.*, 2006; Tir and Ackerman, 2009). Nevertheless, only recent scholarship has investigated the outcomes of institutional cooperation in various sociopolitical contexts. There is a dearth of literature to understand the outcomes of water diplomacy, which represent countries' diverse interests in the sharing of water resources. Questions remain to be answered, such as when and how diplomatic activities are developed to realize countries' interests in the sharing of water resources (Tir and Ackerman, 2009).

## Outcomes of water diplomacy

Diplomatic initiatives may play a significant role in mitigating interstate disputes over water resources. For diplomatic cooperation to succeed, it is necessary to understand its effects. Water diplomacy is affected by multiple factors including varying degrees of dependence over shared water resources and the foreign relations with relevant neighbouring riparian countries. Consequently, the outcomes differ.

### *Trust-building and peacemaking on sharing water*

Trust-building is a principle of great importance in water diplomacy. Indeed, it is based on pleasant foreign relations which are a prerequisite for stable regime formation (Haftendorn, 2000; Wolf *et al.*, 2003; Brochmann and Gleditsch, 2012). However, in water diplomacy, trust-building is often neglected. For example, mistrust often occurs due to the asymmetric possession of information over shared water basins. Therefore, the availability of relevant hydrological data represents an advantage to those states that possess such information (Timmerman and Langaas, 2004; Grossmann, 2006). As analysed above, because of countries' differing geographical locations, asymmetries in the possession of information may constitute a major source of mistrust. Hence, the primary function of a water treaty, such as WMIs, is to serve monitoring purposes that enhance cooperation (including information exchange) between riparian states over the management of shared rivers. In particular, one apparent focus in developing WMIs is to protect citizens when human security is threatened by disastrous events such as droughts and flooding. Most nations recognize their obligation in the protection of individual human security. Almost immediate responses can be provoked by nation states' peacemaking activities (Conca *et al.*, 2006; Karkkainen, 2006).

WMIs developed through information sharing and therefore function as a diplomatic initiative to promote trust-building among riparian states with shared rivers (Conca *et al.*, 2006; Karkkainen, 2006; Zawahri, 2008, 2009). The primary goal of WMIs is to eliminate the mistrust that can occur among neighbouring riparian states (Stinnett and Tir, 2009). It is recognized that water diplomacy serves to resolve specific problems such as overcoming tensions among countries (Brochmann and Hensel, 2009; Bernauer and Kalbhenn, 2010). The smooth exchange of water data promotes communication and understanding between neighbouring riparian states (von Stein, 2008) and also helps to facilitate transparency and mutual assurance (Burton and Molden, 2005; Gerlak *et al.*, 2011).

### *Water management and resource governance*

Clean water is a natural resource of fundamental importance to human society. Nation states are key actors in providing environmental services as public goods. Water management serves multiple objectives and navigates between competing interests (Wolf *et al.*, 2005). Based on successful diplomatic initiatives, substantial collaboration can be developed in various areas of water management including adaptive forms of environmental management (Dietz *et al.*, 2003; Lindemann, 2008; Raadgever *et al.*, 2008) and the monitoring of river basins.

Water diplomacy facilitates improved river-system benefit sharing both within and between countries, particularly in the area of satisfying human needs and interests (see also Sadoff and Grey, 2002, 2005). In both theory and practice, integrated water resource management (IWRM) is believed the best method to seeking benefits from shared water basins (Sadoff and Grey, 2002: 392, 2005; Diez *et al.*, 2003). This approach aims to promote ecosystem improvements or the sustainability of river basins, which is to maintain river ecosystem integrity while also helping to develop economic interests from water resources. It considers river basins as the basic unit in the planning and management of water resources. Public participation in decision-making is viewed as an important principle in water policy. Recently, closely related conceptual tools, such as the Water-Food-Energy-Ecosystems nexus have also been developed as advanced approaches to promoting the sustainable management of water resources (see Moynihan and Magsig, 2014).

Trust-building in water sharing and collaboration in water management represent two distinct outcomes in water diplomacy. Optimistic views regard these two dimensions as a holistic process, with the latter seen as providing the foundation for sustainable water management to be achieved (Haftendorn, 2000; Brochmann and Gleditsch, 2012). However, negotiations to develop collaborative water management are not a linear process. Empirical findings show that successful water institutions are established under certain conditions that allow for diplomatic initiatives to be formed (Bernauer, 1997, 2002; Dinar, 2009), including international factors such as the involvement of international institutions (Salehyan, 2008). This phase sees a broader scale of information being shared and heightened institutional arrangements being made. Then, sustainable water management may represent a further stage when deepened collaboration is developed. Furthermore, case studies have suggested that such negotiation could be a holistic process. In post-conflict situations, water is argued to constitute an important factor for advancing socio-economic development. The restoration and management of water resources is seen to be significant (Montville, 1995; Cuhadar, 2009; Weinthal *et al.*, 2011; Beck, 2015). Hence, some argue that by eliminating mistrust, successful water diplomacy may succeed in promoting socio-economic development in a sustainable fashion (Stinnett and Tir, 2009).

However, with varying incentive structures being key factors underlying negotiations, asymmetric preferences can affect the prospects of developing broad collaboration that promotes sustainable water management (Berardo and Gerlak, 2012). Countries have various incentives to form cooperative institutions (see Stinnett and Tir, 2009), and concerns over water resource management and political interests coexist, resulting in various different outcomes. Regarding water management on the shared international rivers, collective management may become difficult to achieve, although the independent use of water resources within a nation does not necessarily impact on other nations' interests. Some water diplomacy leading to WMIs may be underpinned by an agenda to reap profits from water management (Fischhendler, 2008). Indeed, in situations where political interests are shown, water diplomacy has come to develop a negative character, and indeed has been referred to as disguised cooperation (Dombrowsky, 2003) or a matter of domination (Selby, 2003).

## Water diplomacy and the formation of water management institutions

### *The content of water diplomacy*

States show a continuum of diplomatic commitments over shared water, with the trajectory of their actions to collaborate in water management varying widely. These actions can range from states taking independent actions to comply with water agreements to their developing collaboration and establishing joint actions to address water sharing issues (Sadoff and Grey, 2005). Water management institutions (WMIs) represent arrangements that result from water diplomacy among riparian states.

With regard to the governance structure of WMIs, it requires inclusive participation and equal position among stakeholders. Although national governments are official representatives of negotiations, the inclusion of all nation states with shared water basins is regarded as key to forming effective WMIs. As Schmeier *et al.* (2016) argues, the quality of actors is crucial to the effectiveness of WMIs to promote the management of water basins. Some even argue that the cross-sectoral representation of actors is crucial to the effectiveness of WMIs in promoting the sustainable and equitable management of water basins (Schmeier *et al.*, 2016).

The formation process of WMIs is a challenging process to international relations among riparian states. Scholarship has identified a range of diplomatic mechanisms including both formal and informal approaches that facilitate cooperation in water basin management. Emphasis is placed on formal cooperative arrangements (Hensel *et al.*, 2006). One approach is to set up formal organizations that facilitate negotiations – river basin organizations (RBOs) – as intergovernmental organizations. If an RBO adopts IWRM water management principles, it is likely to incorporate all actors involved (Schmeier *et al.*, 2016). Literature on conflict resolution informs us that another identified mechanism for institutionalizing cooperation is to rely on pre-existing cooperative mechanisms (LeMarquand, 1977; Bernauer, 1997; Wolf *et al.*, 2003;

Conca *et al.*, 2006). Existing channels, particularly those that promote economic cooperation, reduce transaction costs (Tir and Ackerman, 2009: 628) and provide a platform for countries to commit to cooperative water management (Neumayer, 2002; Espey and Towfique, 2004; Dosch and Hensengerth, 2005; Tir and Ackerman, 2009).

Informal diplomatic initiatives may also prove useful. In the literature on water dispute resolution, it is suggested that informal mechanisms facilitate effective cooperative efforts among nations. Uprety and Salman (2011) suggest there is inflexibility in treaties designed to respond to changes in water management and that such treaties or WMIs may result in ineffective conflict resolution. Means of informal cooperation, including collective research or informal networking among policy entrepreneurs or non-state actors, are neglected (Kistin and Ashton, 2008). Flexible WMIs are needed so that they can be effectively implemented. For instance, in the Southern African Development Community, forms of collective research or informal networking among key stakeholders are proving effective in the regulation of transboundary waters, particularly in situations whereupon certainty is difficult to achieve in the management of shared river basins (Kistin and Ashton, 2008). In situations where acute competition over water quantity exists, a certain degree of flexibility contributes to positive cooperative efforts (see Drieschova *et al.*, 2008).

At the same time, the formation of WMIs greatly relies on the capacities of countries. Essentially, if decisions to develop collaborative water management are made, countries will encounter challenges to internationalizing their domestic water policies and this may require a heightened scale of coordination and more complicated governance arrangements (Pahl-Wostl *et al.*, 2012). In some developing countries, the government is incapable of establishing basin-wide cooperation. Assistance from international actors such as intergovernmental organizations and NGOs has proved crucial. Salehyan (2008) has illustrated how in contexts where political instability exists, the instrumental assistance provided by international actors proves effective to promoting treaty formation. International actors provide effective assistance in a range of areas including treaty formation (Salehyan, 2008), the restoration and management of water resources (Weinthal *et al.*, 2011; Beck, 2015), and in reconciliation and peace-making processes (Montville, 1995; Beck, 2015).

### *Water management mechanisms in WMIs*

Acting as mediation bodies in countries' sharing transboundary waters, the scope of institutional arrangements is another heatedly debated issue. Many agree that WMIs' content is more expansive than ordinary state-led cooperation. Recent scholarly work advocate that WMIs are efficient in serving multipurpose aims to achieve goals of integrated water basin management (Dombrowsky, 2007; Sadoff *et al.*, 2008; Schmeier *et al.*, 2016), rather than in focusing on single-issue management upon which disagreements arise (Marty, 2001).

In many existing WMIs, information sharing is adopted as a policy tool. According to scholarship focusing on water agreements, information disclosure represents a key international norm for mitigating conflicts (Mitchell and Zawahri, 2015). Water information, or water data, refers to a broad range of information related to the management of transboundary water basins and may include both scientific and social context data. In particular, information disclosure represents a key international norm highlighted by advocates of liberal international order in global environmental governance (cf. Gupta, 2008). Furthermore, information sharing provides insights on WMIs that potentially serve diversified sectoral interests. Two broad dimensions that are closely interconnected can be identified, by investigating the areas in which and purposes for which information is shared. These are: to promote water management that orients towards protecting human security; and to promote trust-building that orients towards resolving specific causes of conflict in the sharing of transboundary water resources.

In sum, depending on the nature of water diplomacy, their outcomes differ and various WMIs may be consequently formed. The existing institutional cooperation represents practices and normative values found in specific socio-political contexts, particularly those in Western Europe or North America. It is likely that these may not immediately apply to Asia, which may have its own characteristics.

## China's water diplomacy

On the issue of water sharing, China falls into a paradoxical dilemma that shows the complex rationale in its appropriation of natural resources. On the one hand, China willingly and rightfully exercises its sovereign rights to plan and manage resources within its territory, and its upstream location on a number of transboundary rivers places it in an advantageous position to do so. On the other hand, the country is aware of potential tensions this exercising of its rights may bring and has displayed certain degrees of responsibility that show its willingness to accommodate other countries' interests, even when compromise is required. In analysing the nature of substantial water diplomacy that has been or is being formed in Asia, in this section we will explore the potential nature and outcomes of China's water diplomacy. Its key features of WMIs and the role China plays in WMIs will be discussed.

### *Diplomacy and China's strategic security interests*

Despite China's policy advocate, analysis of China's foreign policy agenda informs us that its discourses dynamically evolve. Disagreements exist in the understanding of China's core interests, recognizing that China has shown shifting attitude towards forging cooperation (Hill, 2015). Transboundary water issues constitute one of the key topics in the debate of China's core interests (Zeng *et al.*, 2015). Water is multifaceted issue, and complicated perceptions have been formed.

China's emphasis on water as a security strategy is contended. Securitization over water resources is advocated by the realism prism, a powerful perspective in understanding China's strategic interests over international rivers. The realism perspective suggests that in international relations, water is viewed as a key issue area for China to secure leadership in the region (Liebman, 2005). This view strongly emphasizes the implications of China's water use in relation to its national interests. In particular, this line of analysis argues that water resources are sought in material terms, such as defence interests towards downstream states, acting as water hegemon, which is permitted by China's geographical location (Chellaney, 2011).

However, it remains to be seen how China's security strategy develops. Indeed, domestic water resource management has been an important factor that motivate for the country to develop securitization strategy. For instance, in some cases such as the Mekong River basin, it is recognized that Beijing exhibits an intention to enhance opportunities to develop the West including the provinces Yunnan and Guangxi (Magee, 2012; Biba, 2014). However, it is argued that water as a source of material interest has not become a threat to national security, considering that the country has developed economic prosperity (Biba, 2014).

Water, representing a security concern, is also advocated as non-traditional security (NTS) concern. It is noted that the notion of NTS remains ambiguous. In Chinese scholarship, environmental security remains an ambiguous term. It is argued that environmental security represents non-military security concerns (Yu and Wang, 2006). It poses threats to human security and is separate from a conventional security concern, which is related to the adoption of armed forces (Wang, 2003). However, despite the fact that natural resources, such as water, are high on the national security agenda, they are subordinate to national military security. In April 2014, six broad categories of security were outlined in the first Plenary Session of the National Security Commission.[1] We are reminded that in the official statement, the national security concerns are ordered by their priority on the security agenda (You, 2015). State security (or regime security), sovereignty security (or territorial, maritime and continental securities) and military security remain the core of China's interests. Economic security (or energy and natural resources) ranks fourth, while societal security (or cultural, human and nuclear securities) and other security issues (such as science and technology, health, food and information securities) rank last. It is thus interpreted that China's commitment to realizing water interests is secondary to traditional security concerns, such as the intactness of its political institutions and border integrity. Additionally, it is interpreted that China's intention to protect human security from environmental changes does not necessarily result in placing environmental issues on its security agenda. In the understanding of the notion 'environmental security', emphasis is placed on the related securitization measures. It is pointed out that environmental security is of secondary concern to national security, since it relates to non-militant approach (Xiao, 2007; Zhang, 2012). Therefore, to many, China's perception of water or water-related disasters may contrast significantly to how environmental change is perceived in the Western European context, whereby a lack of sustainability in river basin management is seen as a major concern for human security. According to critical security studies, the environment has been identified as a source of threat, particularly when scientific findings have shown that human vulnerability is strongly associated with environmental changes (Peoples and Vaughan-Williams, 2010). Consequently, national governments in developed societies are likely to respond by highlighting the importance around the issue of water management.

As the country is undergoing evolution from pursuing policy changes in defence to NTS, it is likely that water represents a new arena where unique strategic interests and approaches are displayed in the policy of international water sharing. It has been suggested that the Chinese authorities have shown increasing interest in responding to water-related disasters that pose challenges to human security (Breslin, 2015), including drought, flooding and those that affect public health. Interestingly there have been occasions where Chinese authorities have also shown a sense of responsibility towards the welfare of individuals in other nations (Yeophantong, 2013). Questions have been raised: to what extent water constitutes a key policy in China's foreign policy agenda; and whether the politicization of water would induce the use of military measures or other coercive means. More importantly, what are the key factors that affect Chinese policymakers' politicization of water, considering that water is perceived in various ways?

## *The formation of WMIs and the role of China*

China's image as a superpower has played a misleading role. International relations inform us that sovereign states are expected to act out of self-interest. The concept of a superpower state is, in particular, pertinent in this respect. Such states seem to be able to regularly escape the constraints of international politico-economic structures. In other words, superpowers seem to enjoy a degree of agency that many smaller nations lack. In addition, superpowers are seen as both those which truly can act out of self-interest (as they do not have to fear repercussions) and as those which need to be particularly responsible in their behaviour. Greater expectations often exist regarding a state's virtuousness, the more powerful the state in question is (Gilley, 2013). This leads to a paradoxical situation in which the mere fact that a state is more powerful than another or geopolitically better situated is seen as a reason for that state to engage in collaborative practices and to let others share in its benefits. For China, this can complicate its quest for smooth negotiations with its riparian neighbours.

We are reminded that contextual conditions exist to establish strong institutional mechanisms that can successfully coordinate countries' interests in water. In Asia, where regional interests remain uncoordinated (Mirumachi, 2015), diplomatic initiatives may be formed in a fashion that bears its own character. If a multilateral mechanism is to be developed, it is likely to display a regional feature. Scholarship informs us that regionalism is likely to be powerful in the formation of regional mechanisms (Wouters, 2002). Asian countries in particular show a strong likelihood of forming a regionalist perspective towards sovereignty (Acharya, 2009). The reason may lie in these countries' self-protection from more powerful nations. Therefore, the scope of actors to include in WMIs remains a key question. In particular, China has shown a strong position on who is deemed beneficial to WMIs. A regionalized approach is set out in its agenda that excludes international actors from dealing with regional disputes (Li, 2011). It is not clear whether China would engage with international norms and ethical codes in the sharing of water resources.

Asian states in particular have displayed an inclination to develop informal diplomacy. Such practices are considered beneficial to region-building initiatives (Elliott, 2012). China has also shown cultural features in adopting informal networks and personal connections in its social and political life. In various aspects of environmental governance, it has been identified that such practices are adopted in various sectors, including policy processes and the development of environmental activism (Xie, 2009). Therefore, possibilities exist for informal mechanisms to play an active role in coordinating national interests in the cooperative management of shared rivers. Again, this reminds us that Asia may show unique features and opportunities for forming cooperative partnerships.

### *Outcomes of China's water diplomacy*

Two interlinked dimensions are identified regarding China's behaviour over shared water resources (Table 4.1). The combined effects of these two sets of criteria have the implications of maintaining regional order. The vertical dimension discusses the extent to which China combines its diplomacy initiatives to promote water-basin management. It outlines that various interests may be vested in China's intentions to form institutional arrangements with its riparian neighbours. The nature of such commitments to water diplomacy can be distinguished from strategic interests that are likely to neglect the benefits that China can gain from basin-wide sustainable water management. The horizontal dimension assesses the level of institutionalization of WMIs that have

*Table 4.1* Scenarios of China's behaviour and its implications for promoting basin-wide diplomacy

| | *Low level of institutionalization* | *High level of institutionalization* |
|---|---|---|
| Low level of cooperation in water-basin collaboration | A | B |
| High level of cooperation in water-basin collaboration | C | D |

been formed. China may be incentivized to participate in institutional agreements. However, such initiatives may not necessarily lead to substantial cooperation over transboundary interests. The outcomes depend greatly on the negotiations with downstream states. Therefore, there are three scenarios that reflect China's intentions and the result of institutional cooperation from its diplomatic moves (scenario C is not a valid situation in this case). This model provides a framework of analysis indicating the nature of Chinese diplomacy and the outcomes of its contributions to institutional cooperation over water resource management.

Scenario A is likely to be the worst situation for both China and its neighbouring riparian states. China is likely to adopt a strategic approach that prioritizes national interests in water management. Accordingly, its diplomatic initiatives are likely to be a symbolic gesture that takes little consideration of water resource management. Such WMIs may potentially endanger China's foreign relations with its neighbouring states, and may lead to political tensions. Hence the country's reliance on water sharing for security interests may fail.

Scenario B describes China's attitude in progressing its sovereignty rights with little consideration of strategic interests in this process. It represents a situation of how China will probably behave when faced with transboundary environmental issues in its intention to benefit from basin-wide water diplomacy. However, although the country may be keen to develop a high level of institutionalized water diplomacy, as a result of lacking persuasion with regional neighbours, disagreements are likely to be raised when negotiating individuals' interests over collaborative water management.

Scenario D is the best scenario for China as well as for all other nations involved. It is foreseeable that China takes a proactive position in promoting sustainable water resource development on a regional scale. China has developed a broader consideration of basin-wide interests while being successful in developing institutional agreements with other nations. Being aware of its regional status as well as its advantageous geographical location, it is likely to commit to a high level of institutional cooperation. In comparison to scenario A, this situation probably also proves to be the most challenging, as the country would play a leadership role. As outlined earlier, being a responsible actor may require that China take a leading role, both in material assistance and policy transfer.

## Methodology

### *Advantages of case-study research*

This study examines China's multifaceted water policies and outcomes managing international rivers. Such policies are heavily influenced by bilateral relations and competing needs in water resource management between China and its neighbouring states. It is true that scholarship has already examined specific transboundary water resource management using the case-study methodology. It adopts case-study research methods to provide a critical analysis of the effectiveness of different cooperative initiatives over water disputes between China and its neighbouring states.

The case-study methodology is used for two reasons. First, according to Yin, case-study research is especially relevant if 'how' and 'why' questions are asked about a contemporary set of events, over which the investigators have little or no control (Yin, 1984). This method enables us to investigate the development of water vows representing varying national interests that are vested in the resolution of water disputes, such as interests in water resource management, economic profits and strategic interests. Case-study research can describe and explain the complex and entangled patterns, structures or processes (Verschuren, 2003). It is therefore suited to studying the formation of cooperative initiatives between China and other countries that represent dynamics in relation to varying contexts where river characters differ.

Second, qualitative case-study research forms a suitable strategy to understand the levels of cooperation between countries with water resource interactions. It has the potential to produce strong evidence in exploring the relationship between these countries' varying interests and efforts to form institutional agreements. Hence, the findings can provide rich insights into the various outcomes of cooperative initiatives aimed at solving water disputes in Asia.

However, while providing in-depth understanding of particular cases, this methodology also has its limitations. By utilizing the case study as a research methodology, the research has both ambiguities in its internal and external validity (Verschuren, 2003). Externally, this study may be less powerful than quantitative methods for providing generalized theoretical propositions. What can be learned from this study will primarily relate to the specific water resolutions found between China and the neighbouring states with whom it shares water resources. Yet, by exploring four cases, this study represents a variety of local conditions in river characteristics and key non-water issues that are identified as affecting foreign relations, especially in South Asia and South-East Asia where many large international rivers are found. Therefore, by utilizing multiple cases, this study has much potential to contribute to theoretical refinement on institutional cooperation and conflict resolution over shared water resources.

This study also has limitations regarding its internal validity, which is researcher-dependent (Creswell, 1994; Verschuren, 2003). This weakness is to some extent avoidable as the outcomes of interstate water disputes are mostly accessible through publicly available sources such as government documents and news articles. On rare occasions, it is likely that interstate interactions are secretly conducted. Hence, this study is likely to avoid any major internal validity issues.

### *Selection of case studies*

The study will undertake a range of case studies of China's water policy as it is applied to various international rivers and neighbouring countries: these range, geographically, from South-East Asia (ASEAN countries) to South Asia (Bangladesh and India), Central Asia (Kazakhstan) and North-East Asia (Mongolia and Russia). In each of these chapters, the security of major international rivers will be examined by assessing the intensity of conflicts rooted in issues around how water resources are managed, the performance of relevant environmental agreements, and efforts at water diplomacy.

The research involves case studies of China's water diplomacy over four transboundary river basins. Each case begins with an overview of the river basin, including the features of the river, the level of water insecurity China and its key riparian neighbour have shown, and China's position and policies over the management of the rivers. For each case, we then investigate the discourse of disputes over the shared rivers before analysing the rationale and mechanisms of China's diplomatic initiatives aimed at mitigating the disputes. Sub-cases are selected, consisting of countries that China shares rivers with and with which disputes have developed. In this study, four international rivers are selected as cases to examine China's diplomacy and its impact on water disputes, namely the Ganges–Brahmaputra–Meghna (GBM), the Mekong, the Ili and the Amur Rivers.

For most of the rivers, the sub-cases involve one country that is located immediately downstream on the shared river system. In the case of the Mekong River where the institutional water body the Mekong River Commission has already been established, it is adopted as a sub-case to examine Chinese diplomacy and its effects on water disputes.

Three criteria have been applied in selecting these cases, namely asymmetries in the countries' security, the nature of countries' political system, and differences in their economic growth (see Table 4.2). One key factor that impacts on the countries' water diplomacy is the level of asymmetries in states' level of insecurity over river basins. In this study, the water resources exploitation rate is adopted to measure the countries' varying degrees of reliance on the water basin. The water resource exploitation rate refers to the ratio between a country's water extraction and the total water resources. This approach considers water uses which are to be divided between exploitation for economic resources and preservation to restore a river's ecological environment. It is generally assumed that a country is experiencing insecurity if its water resource exploitation rate reaches 40 per cent.

Such asymmetries are likely to result in different diplomatic initiatives that may succeed differently in the forming of collaborative water resource management. Based on a study of original panel data on WMIs, Ovodenko (2014) argues that countries' symmetrical interdependence on rivers represents a situation where integrated management is likely to be formed. This is because the countries involved are more likely to develop congruent preferences over the form of institutional arrangements (Dieperink, 2011; Zawahri and Mitchell, 2011).

In the selection of cases, varying degrees of asymmetries have been found between China and its riparian neighbours, ranging from low to high levels of asymmetries (Table 4.3). More specifically, high levels of asymmetries exist in

*Table 4.2* Contextual factors in the river system case studies

| | GBM *River basin (with India)* | *Mekong River basin (with the LMCs)* | *Ili River basin (with Kazakhstan)* | *Irtysh River basin (with Kazakhstan)* | *Amur River basin (with Russia)* |
|---|---|---|---|---|---|
| Water insecurity | Large difference[2] | Small difference | Small difference | Small difference, both threatened | Large difference |
| Status of security | China secured; India threatened | Both secured | Both threatened | Both relatively secured | China threatened; Russia secured |
| Regime type | Democratic (India) | Democratic (riparian states) | Democratic (Kazakhstan) | Democratic | Democratic |
| Economic growth | Very low | Medium-low | Very high | Very high | Very high |

Notes

1 LMCs = Lower Mekong Countries.

2 India has displayed a high ratio of water resource exploitation rate on the Ganges and the Indus Rivers. Although it is secured on the Yarlung Zangbo River, as the GBM is a major source of water for India, we can infer that the country is threatened on the GBM.

*Table 4.3* Asymmetries in countries' water resource exploitation rates

| | *China (%)* | *Riparian country and its water resource exploitation rate (%)* | |
|---|---|---|---|
| GMB River basin | 0.7 | India | 2.4 |
| Mekong River basin | 3.7 | The highest being Vietnam | 14.9 |
| | | The lowest being Laos | 1.2 |
| Ili River basin | 36.2 | Kazakhstan | 42.7 |
| Irtysh River basin | 15.3 | Kazakhstan | 12.8 |
| Amur River basin | 40.5 | Russia | 0.9 |

Source: Yan (forthcoming).

the sharing of the Amur River, and in the case of the Mekong and the Ili Rivers. Although India has shown a low degree of reliance on the Yarlung Zangbo River, its reliance on the Indus and Ganges Rivers are high, hence it has a high degree of insecurity.

The nature of the governing regime is a second contextual factor that may affect the outcomes of water diplomacy as countries' behaviours in forming cooperative partnerships may differ. It is contended that the nature of the countries' political systems affects the approaches that authorities may take in dealing with water disputes, such as the compliance with global water norms, or by incorporating the principles of water rights. Gerlak *et al.* (2011) suggest that authoritarian states, compared with democratic systems, are unlikely to join cooperative mechanisms in water disputes and collaborative water resource management. Hence, these states are more likely to develop intense political tensions with other riparian states (Stinnett and Tir, 2009).

There is a significant difference between the political systems active in the four cases; namely that India, Kazakhstan, Russia and some of the riparian countries in the Lancang-Mekong River basin are democracies. However, it should be noted that these states have short democratic histories and there are shortfalls in the democracies operating in these states. Overall, there is an absence of a regional commitment to establish strong democratic communities (Kraft, 2006). As for their participation in multilateral water agreements, only Kazakhstan and Russia are parties, indicating their stronger commitment to international cooperation. Therefore, across the cases there are varying stances on international cooperation that may affect the development of Chinese diplomacy and the outcomes of any cooperative initiatives.

Third, economic development level is another factor that impacts on the nature and outcomes of China's water diplomacy. Economic conditions (calculated by GDP and GDP per capita) affect the countries' formation of coordinated interests in water. This can be a potential source of conflict. Determined by a country's GDP level, each country has demonstrated varying degrees of preference to exploit and manage water resources (Elhance, 1999; Dinar *et al.*, 2010). It is contended that high GDP countries are more capable of exploiting economic interests in water, which may lead to increased competition over resource extraction. Hence, the unequal economic growth in different regions may result in increased tensions among nations when negotiating water institutions (Elhance, 1999).

Additionally, water management is costly and requires high degree of material power, including finance and know-how. Hydropower development, for instance, requires a large amount of up-front investment. Human resources and expertise are required to manage its operation. And the capacity of exploiting shared water resources represents a challenge to a number of states in the chosen sub-cases. Those that lack material capacity to exploit shared water resources are likely to become disadvantaged. Considering the weak water management capacity of many riparian nations, a responsible state would have to engage in capacity building that facilitates the inclusion of its poorer riparian partners. Needless to say, these issues pose real challenges for China to develop successful water diplomacy.

### *Methods for data collection*

This project employs multiple research methods to generate a rich array of data to understand interstate interactions over shared water. Triangulation of the data has been achieved through the combination of quantitative and qualitative methods and the adoption of a spatial vulnerability approach. The qualitative methods include in-depth, one-on-one semi-structured interviews and focus group discussions with officials in relevant government departments and other water resource scientists and experts; analysis of archived news reports on regional water disputes; and analysis of other discourses surrounding the issue of water conflicts.

- In-depth interviews

  In order to explore the extent to which cooperative institutions are perceived to contribute to solving problems identified in the sharing of water resources, the main targets for the in-depth interviews were influential elites, experts and policymakers.

  One of the authors works closely with the Chinese Ministry of Water Resources and was involved in consultancy on the management of a trans-boundary water management project. Accordingly, first-hand knowledge of Chinese policy development on water resources was gained for this study. Semi-structured interviews were conducted with the national government and relevant NGOs in both India and China. Major topics of discussion were their concerns over water resource management, and other issues surrounding water information sharing agreements and how they are complied with.
- Archival studies

  Through archival analysis, government documents are reviewed to understand the formal processes and content of water diplomacy mechanisms established among neighbouring riparian states. Openly published documents are accessed from government websites. This study carefully examines the discourses surrounding the issue of water disputes on the case study river basins. A particular focus is on materials generated by disputes or claims from all sides since 2000, when China began to unilaterally develop upstream projects. Such information can be found in news reports and policy papers published since the establishment of formal and informal mechanisms of cooperation.

- Geographical modelling
  This book will also incorporate findings on water security of Asia's major international rivers from an independent project led by Shaofeng Jia and Jiabao Yan. Modelling using a Soil and Water Assessment Tool is the main approach adopted to investigate the vulnerability displayed by China and its riparian neighbours on the river basins including the GBM, Lancang-Mekong, Ili and Amur Rivers. The specifics can be found in the doctoral thesis of Jiabao Yan (Yan, forthcoming).

## Note

1 China Daily, 16 April, 2014. Available at: http://cpc.people.com.cn/n/2014/0416/c64094-24900492.html (accessed 30 August 2017).

## References

Acharya, A. 2009. *Whose Ideas Matter? Agency and Power in Asian Regionalism*, Ithaca, NY: Cornell University Press.

Amacker, C. 2011. The functions of diplomacy, *E-international Relations*. Available at: www.e-ir.info/2011/07/20/the-functions-of-diplomacy/ (accessed 11 July 2017).

Barston, R.P. 1988. *Modern Diplomacy*, London: Longman.

Beck, A. 2015. The UN Peacebuilding Commission and the potential of water in post-conflict development, governance and reconciliation, *Water International*, 40(2): 215–230.

Berardo, R. and Gerlak, A.K. 2012. Conflict and cooperation along international rivers: crafting a model of institutional effectiveness, *Global Environmental Politics*, 12(1): 101–120.

Bernauer, T. 1997. Managing international rivers, in O. Young, ed., *Global Governance: Drawing Insights From the Environmental Experience*, Cambridge, MA: MIT Press, pp. 155–195.

Bernauer, T. 2002. Explaining success and failure in international river management, *Aquatic Science*, 64(1): 1–19.

Bernauer, T. and Kalbhenn, A. 2010. The politics of international freshwater resources, in R.A. Denemark, ed., *The International Studies Encyclopedia*, Malden, MA: Wiley-Blackwell, pp. 5800–5821.

Biba, S. 2014. Desecuritization in China's behavior towards its transboundary rivers: The Mekong River, the Brahmaputra River, and the Irtysh and Ili Rivers, *Journal of Contemporary China*, 23(85): 21–43.

Breslin, S. 2015. Debating human security in China: Towards discursive power? *Journal of Contemporary Asia*, 45(2): 243–265.

Brochmann, M. and Gleditsch, N. 2012. Shared rivers and conflict: A reconsideration, *Political Geography*, 31: 519–527.

Brochmann, M. and Hensel, P. 2009. Peaceful management of international river claims, *International Negotiation*, 14(2): 391–416.

Burton, M. and Molden, D. 2005. Making sound decisions: information needs for basin management, in M. Svendsen, ed., *Irrigation and River Basin Management: Options for Governance and Institutions*, Wallingford: CABI, pp. 51–74.

Chellaney, B. 2011. *Water: Asia's New Battleground*, Washington, DC: Georgetown University Press.

Conca, K., Wu, F. and Mei, C. 2006. Global regime formation or institution building? The principled content of international river agreements, *International Studies Quarterly*, 50(2): 263–285.

Creswell, J.W. 1994. *Research Design: Qualitative and Quantitative Approaches*, Thousand Oaks, CA: Sage.

Cuhadar, E. 2009. Assessing transfer from track two diplomacy: The cases of water and Jerusalem, *Journal of Peace Research*, 46(5): 641–658.

Daoudy, M. 2007. Benefit-sharing as a tool for conflict transformation: Applying the Inter-SEDE model to the Euphrates and Tigris river basins, *The Economics of Peace and Security*, 2(2): 26–32.

Dieperink, C. 2011. International water negotiations under asymmetry: Lessons from the Rhine chlorides dispute settlement (1931–2004), *International Environmental Agreements*, 11(2): 139–157.

Dietz, T., Ostrom, E. and Stern, P.C. 2003. The struggle to govern the commons, *Science*, 302(5652): 1907–1912.

Dinar, A., Blankespoor, B., Dinar, S. and Kurukulasuriya, P. 2010. Does precipitation and runoff variability affect treaty cooperation between states sharing international bilateral rives? *Ecological Economics*, 69(12): 2568–2581.

Dinar, S. 2009. Scarcity and cooperation along international rivers, *Global Environmental Politics*, 9(1): 109–135.

Dombrowsky, I. 2003. Water accords in the Middle East peace process: Moving towards cooperation? In H.G. Brauch, P.H. Liotta, A. Marquina, P.F. Rogers and M.E.-S. Selim, eds, *Security and Environment in the Mediterranean: Conceptualising Security and Environmental Conflicts*, Berlin: Springer, pp. 729–744.

Dombrowsky, I. 2007. *Conflict, Cooperation and Institutions in International Water Management: An Economic Analysis*, Cheltenham: Edward Elgar.

Dosch, J. and Hensengerth, O. 2005. Sub-regional cooperation in Southeast Asia: The Mekong Basin, *European Journal of East Asian Studies*, 4(2): 263–286.

Drieschova, A., Giordano, M. and Fischhendler, I. 2008. Governance mechanisms to address flow variability in water treaties, *Global Environmental Change*, 18(2): 285–295.

Elhance, A.P. 1999. *Hydropolitics in the Third World: Conflict and Cooperation in International River Basins*, Washington, DC: US Institute of Peace Press.

Elliott, L. 2012. ASEAN and environmental governance: Strategies of regionalism in Southeast Asia, *Global Environmental Politics*, 12(3): 38–57.

Espey, M. and Towfique, B. 2004. International bilateral water treaty formation, *Water Resources Research*, 40(5): 201–222.

Fischhendler, I. 2008. Ambiguity in transboundary environmental dispute resolution: The Israel–Jordanian water agreement, *Journal of Peace Research*, 45(1): 91–110.

Fox, C. and Sneddon, C. 2007. Transboundary river basin agreements in the Mekong and Zambezi basins: Enhancing environmental security or securitising the environment, *International Environmental Agreements: Politics, Law and Economics*, 7(3): 237–261.

Gerlak, A.K., Lautze, J. and Giordano, M. 2011. Water resources data and information exchange in transboundary water treaties, *International Environmental Agreements: Politics, Law and Economics*, 11(2): 179–199.

Gilley, B. 2013. *The Right to Rule: How States Win and Lose Legitimacy*, New York: Columbia University Press.

Gleick, P. 1990. Environment, resources, and international security and politics, in E. Arnett, ed., *Science and International Security: Responding to a Changing World*, Washington, DC: American Association for the Advancement of Science Press, pp. 501–523.

Grossmann, M. 2006. Cooperation on Africa's international waterbodies: Information needs and the role of information-sharing, in W. Scheumann and S. Neubert, eds, *Transboundary Water Management in Africa: Challenges for Development Cooperation*, Bonn: German Development Institute, pp. 173–235.

Gupta, A. 2008. Transparency under scrutiny: Information disclosure in global environmental governance, *Global Environmental Politics*, 8(2): 1–7.

Haftendorn, H. 2000. Water and international conflict, *Third World Quarterly*, 21(1): 51–68.

Hensel, P., Mitchell, S. and Sowers, T. 2006. Conflict management of riparian disputes, *Political Geography*, 25(4): 383–411.

Hill, D.P. 2015. Where hawks dwell on water and bankers build power poles: Transboundary waters, environmental security and the frontiers of neoliberalism, *Strategic Analysis*, 39(6): 729–743.

Hirsch, P. and Mørck Jensen, K. 2006. *National Interests and Transboundary Water Governance in the Mekong*, Sydney: Australian Mekong Resource Centre, University of Sydney. Available at: http://sydney.edu.au/mekong/documents/mekwatgov_mainreport.pdf (accessed 11 July 2017).

Islam, S. and Susskind, L.E. 2012. *Water Diplomacy: A Negotiated Approach to Managing Complex Water Networks*, New York: Routledge.

Karkkainen, B.C. 2006. Managing transboundary aquatic ecosystems: Lessons from the Great Lakes, *Pacific McGeorge Global Business & Development Law Journal*, 19(1): 209–240.

Kistin, E.J. and Ashton, P.J. 2008. Adapting to change in transboundary rivers: An analysis of treaty flexibility on the Orange-Senqu River basin, *Water Resources Development*, 24(3): 385–400.

Kraft, H.J.S. 2006. *China and Democracy in the Asia Pacific*, UNISCI Dissertation Paper No. 11. Available at: www.ucm.es/data/cont/media/www/pag-72530/UNISCI11Kraft.pdf (accessed 11 July 2017).

LeMarquand, D.G. 1977. *International Rivers: The Politics of Cooperation*, Vancouver: Westway Research Centre, University of British Columbia.

Leonard, M. 2002. *Public Diplomacy*, London: Foreign Policy Centre.

Li, Z. 2011. The issue of cross-border rivers and the relationship between China and the circumjacent countries [Kuaguojie heliu wenti yu zhongguo zhoubian guanxi], *Academic Exploration*, 1.

Liebman, A. 2005. Trickle-down hegemony? China's 'peaceful rise' and dam building on the Mekong, *Contemporary Southeast Asia: A Journal of International and Strategic Affairs*, 27(2): 281–304.

Lindemann, S. 2008. Understanding water regime formation: A research framework with lessons from Europe, *Global Environmental Politics*, 8(4): 117–140.

Magee, D. 2012. The dragon upstream: China's role in Lancang-Mekong development. In J. Öjendal, S. Hansson and S. Hellberg, eds, *Politics and Development in a Transboundary Watershed*, New York: Springer, pp. 171–193.

Marty, F. 2001. *Managing International Rivers: Problems, Politics and Institutions*, Bern: Peter Lang.

Mirumachi, N. 2015. *Transboundary Water Politics in the Developing World*, Abingdon: Routledge.

Mitchell, S.M. and Hensel, P.R. 2007. International institutions and compliance with agreements, *American Journal of Political Science*, 51(4): 721–737.

Mitchell, S.M. and Zawahri, N.A. 2015. The effectiveness of treaty design in addressing water disputes, *Journal of Peace Research*, 52(2): 187–200.

Molle, F., Lebel, L. and Foran, T. 2009. Contested Mekong waterscapes: Where to next? In F. Molle, T. Foran, and M. Käkönen, eds, *Contested Waterscapes in the Mekong Region: Hydropower, Livelihoods and Governance*, Abingdon: Earthscan, pp. 383–405.

Montville, J.V. 1995. The arrow and the olive branch: A case for track two diplomacy, in J.W. McDonald Jr and D.B. Bendahmane, eds, *Conflict Resolution: Track Two Diplomacy*, Washington, DC: IMTD, pp. 7–25.

Moynihan, R. and Magsig, B.-O. 2014. The rising role of regional approaches in international water law: Lessons from the UNECE water regime and Himalayan Asia for strengthening transboundary water cooperation, *Review of European, Comparative & International Environmental Law*, 23(1): 43–58.

Neumayer, E. 2002. Does trade openness promote multilateral environmental cooperation? *World Economy*, 25(6): 815–832.

Ovodenko, A. 2014. Regional water cooperation: Creating incentives for integrated management, *Journal of Conflict Resolution*, 60(6): 1071–1098.

Pahl-Wostl, C., Lebel, L., Knieper, C. and Nikitina, E. 2012. From applying panaceas to mastering complexity: Toward adaptive water governance in river basins, *Environmental Science & Policy*, 23: 24–34.

Peoples, C. and Vaughan-Williams, N. 2010. *Critical Security Studies: An Introduction*, Abingdon: Routledge.

Raadgever, G.T., Mostert, E., Kranz, N., Interwies, E. and Timmerman, J.G. 2008. Assessing management regimes in transboundary river basins: Do they support adaptive management? *Ecology and Society*, 13(1): art. 14. Available at: www.ecologyand society.org/vol13/iss1/art14/ (accessed 10 July 2017).

Sadoff, C. and Grey, D. 2002. Beyond the river: the benefits of cooperation on international rivers, *Water Policy*, 4(5): 389–403.

Sadoff, C. and Grey, D. 2005. Cooperation on international rivers: A continuum for securing and sharing benefits, *Water International*, 30(4): 420–427.

Sadoff, C., Greiber, T., Smith, M. and Bergkamp, G. 2008. *Share Managing Water Across Boundaries*, Gland: IUCN.

Salehyan, I. 2008. From climate change to conflict? No consensus yet, *Journal of Peace Research*, 45(3): 315–326.

Schmeier, S., Gerlak, A.K. and Blumstein, S. 2016. Clearing the muddy waters of shared watercourses governance: Conceptualizing international River Basin Organizations, *International Environmental Agreements: Politics, Law and Economics*, 16(4): 597–619.

Selby, J. 2003. *Water, Power and Politics in the Middle East: The Other Israeli-Palestinian Conflict*, London: I.B. Tauris.

Sosland, J.K. 2007. *Cooperating Rivals: The Riparian Politics of the Jordan River Basin*, Albany, NY: State University of New York Press.

Stinnett, D.M. and Tir, J. 2009. The institutionalization of river treaties, *International Negotiation*, 14(2): 229–251.

Timmerman, J.G. and Langaas, S. 2004. *Environmental Information in European Transboundary Water Management*, London: IWA Publishing.

Timmerman, J.G. and Langaas, S. 2005. Water information: What is it good for? The use of information in transboundary water management, *Regional Environmental Change*, 5(4): 177–187.

Tir, J. and Ackerman, J.T. 2009. Politics of formalized river cooperation, *Journal of Peace Research*, 46(5): 623–640.

Uprety, K. and Salman, S.M.A. 2011. Legal aspects of sharing and management of transboundary water in South Asia: Preventing conflicts and promoting cooperation, *Hydrological Sciences Journal*, 56(4): 641–661.

Verschuren, P. 2003. Case study as a research strategy: Some ambiguities and opportunities, *International Journal of Social Research Methodology*, 6(2): 121–139.

Von Stein, J. 2008. The international law and politics of climate change: Ratification of the United Nations Framework Convention and the Kyoto Protocol, *The Journal of Conflict Resolution*, 52(2): 243–268.

Wang, Y. 2003. Place importance on research of non-traditional security, *Gazette of Chinese Academy of Social Sciences*, 27 May.

Waterbury, J. 1997. Between unilateralism and comprehensive accords: Modest steps toward cooperation in international river basins, *Water Resources Development*, 13(3): 279–289.

Waterbury, J. 2002. *The Nile Basin: National Determinants of Collective Action*, New Haven, CT: Yale University Press.

Weinthal, E., Troell, J. and Nakayama, M. 2011. *Water and Post-Conflict Peacebuilding*, Abingdon: Earthscan.

Wolf, A.T., Kramer, A., Carius, A. and Dabelko, G.D. 2005. Managing water conflict and cooperation, in A.T. Wolf, ed., *State of the World: Redefining Global Security*, Washington, DC: Worldwatch Institute.

Wolf, A.T., Yoffe, S.B. and Giordano, M. 2003. International waters: Identifying basins at risk, *Water Policy*, 5(1): 29–60.

Wouters, P. 2002. Universal and regional approaches to resolving international water disputes: what lessons learned from State practice? The Permanent Court of Arbitration Peace Palace Papers, October: 111–154.

Xiao, R. 2007. Yanjiu he lijie Zhongguo de guoji zeren [Studies and understandings of China's international responsibility], *Shehui kexue* [*Journal of Social Sciences*], 12: 27.

Xie, L. 2009. *Environmental Activism in China*, Abingdon: Routledge.

Yan, J. Forthcoming. Assessment of China's transboundary water resources and water use, doctoral thesis, Chinese Academy of Sciences, Beijing.

Yeophantong, P. 2013. Governing the world: China's evolving conceptions of responsibility, *The Chinese Journal of International Politics*, 6: 329–364.

Yin, R.K. 1984. *Case Study Research: Design and Methods*, Newbury Park, CA: Sage.

You, J. 2015. China's National Security Council: Evolution, rationality and operations, *Journal of Contemporary China*, 25(96): 178–196.

Young, O. 1989. The politics of international regime formation: Managing natural resources and the environment, *International Organization*, 43(3): 349–375.

Yu, X., Pan, Y. and Wang, J. 2006. *Theory of Non-traditional Security* [*Feichuantong anquan gailun*], Zhejiang: People's Press.

Zawahri, N.A. 2008. Designing river commissions to implement treaties and manage water disputes: The story of the Joint Water Committee and Permanent Indus Commission, *Water International*, 33(4): 464–474.

Zawahri, N.A. 2009. Third party mediation of international river disputes, *International Negotiation*, 14(2): 281–310.

Zawahri, N.A. and Mitchell, S.M. 2011. Fragmented governance of international rivers: Negotiating bilateral versus multilateral treaties, *International Studies Quarterly*, 55(3): 835–858.

Zeitoun, M. and Allan, T. 2008. Applying hegemony and power theory to transboundary water analysis, *Water Policy*, 10(S2): 3–12.

Zeng, J., Xiao, Y. and Breslin, S. 2015. Securing China's core interests: The state of the debate in China, *International Affairs*, 91(2): 245–266.

Zhang, W. 2012. Epilogue, in X. Yu, H. Zhu and L. Xu, eds, *Report on China's Non-traditional Security Studies* [*Zhongguo feichuantong anquan yanjiubaogao*], Beijing: Social Sciences Academic Press.

# 5 South-East Asia

## China's water diplomacy on the Mekong River

### 1 Introduction

The Mekong River, known as the Lancang in China, is at the heart and soul of mainland South-East Asia. It traverses the six riparian countries of the Greater Mekong Subregion (GMS), comprising Vietnam, Cambodia, Lao People's Democratic Republic (henceforth Laos), Thailand, Myanmar and China (see Figure 5.1). Some 70 million inhabitants reside within the boundaries of the Mekong River basin. These people depend on the river and its tributaries for food, water, transport and many other aspects of their daily lives, through both large- and small-scale agricultural and horticultural pursuits. The river system is rich in native flora and fauna, with many species being endemic to this unique ecosystem. The Mekong River serves as a symbolic bond of kinship and friendship for the various ethnic groups living within its watershed.

Increasing tensions have arisen among countries that share the Mekong River. China, located on the river's upstream, is facing mistrust from downstream states over its uses of the river. China has been active in initiating projects that impact on the use of the Mekong's water resources, such as dam construction and navigation projects. The country persists in exercising its sovereign rights over the Mekong River and has only formed limited agreements with selected riparian states (see Table 5.1). As China does not consult before initiating its dam projects, it has been criticized for making unilateral decisions (Liebman, 2005; Zhang *et al.*, 2016). The country is assumed to be in a strong position to exert its military powers over the Lower Mekong Countries (LMCs: Vietnam, Cambodia, Laos, Thailand and Myanmar) through flooding, given that the capacity of the dams can directly impact disastrous events in the LMCs.

The political power struggles in South-East Asia have further reinforced mistrust among the LMCs, all of which are strategic collaborators through the Association of Southeast Asian Nations (ASEAN).[1] The Sino-ASEAN relationship is at the centre of controversy. China, owing to its economic power and strategic interests in the region, plays a significant role in South-East Asia. Contention exists among its neighbours regarding China's intention in the region as well as how the country would act to realize its interests (Goh, 2014). At the same time, China's relationships with individual Mekong riparian states

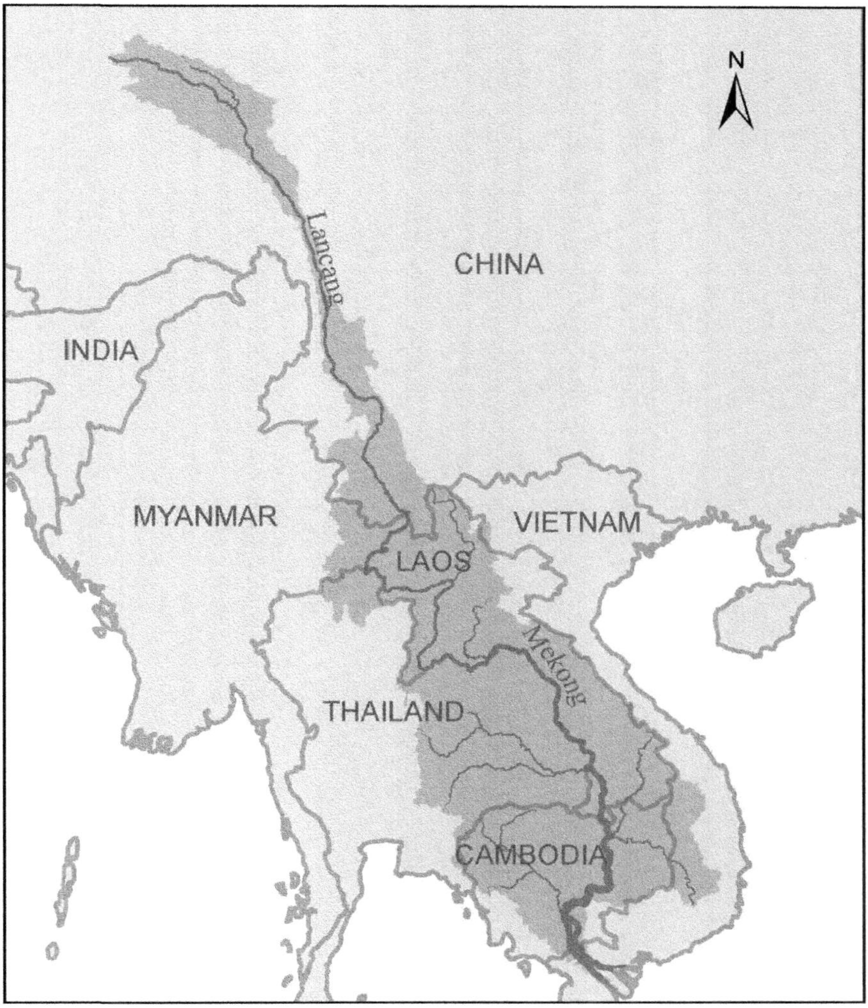


*Figure 5.1* The Mekong River basin.

differ. It is noted that with the diversity of its member states, the individual states and the ASEAN organization have developed incoherent relations and engagements with China (Hernandez, 2008). This has further complicated the interactions with China on the management of the Mekong River and has resulted in various implications for the river's management. Additionally, the increasing tensions in the South China Sea complicate the understanding of the Mekong River system as a holistic system, which is crucial in investigating water politics in this region. Territorial disputes directly impact on China's

*Table 5.1* Institutional arrangements between China and riparian states over the Mekong River

| *Name* | *Time* | *Member* | *Main content* |
|---|---|---|---|
| Framework Agreement on Joint Development of Hydroelectric Resource in Myanmar signed by the Chinese National Energy Administration and the Myanmar Ministry of Electric Power[2] | 1996 and 2009 | China and Myanmar | China participates in hydroelectric cooperation in Myanmar through the pattern of built-operation and transfer (BOT) |
| Agreement on navigation and transport on the Mekong River | 1994, renewed in 2011 and 2015 | China and Laos, and Myanmar and Thailand, respectively | In relation to commercial freight and passenger transport on the Mekong River |
| Agreement on hydro data disclosure | 2010 | China and the Mekong River Commission | China is to share hydro info during the period from 15 June to 15 October every year |
| Agreement on cooperation between China and the Greater Mekong Subregion Mechanism (GMCM) | 2002, renewed in 2005 | China, Myanmar, Laos, Thailand, Cambodia and Vietnam | China and the GMCM cooperates on regional economic integration programmes |
| The Lancang–Mekong Cooperation Mechanism (LMCM) | 2015 | China, Myanmar, Laos, Thailand, Cambodia and Vietnam | Members are to address regional development of the river basins, including through cooperation on interconnectivity, production capacity, cross-border economic development, and agriculture and poverty reduction |

relationship with those that oppose its claiming of territory, leading to a focal point where frictions and clashes are likely to escalate.

This chapter provides a nuanced perspective of China's role and positioning in the region through its focus on water diplomacy and the underlying rationale. The chapter is structured as follows. The next, second, section investigates the characteristics of the Mekong River and analyses the water security of its individual riparian states. The third section reviews China's principles for the planning and management of the Mekong's water resources. In the fourth section, we examine and evaluate how such interests have been vested in the Chinese government's diplomatic initiatives. The last section summarizes China's interests over the Mekong River and provides preliminary conclusions of the role it plays in the management of the river.

## The character and security of the Mekong River basin

The Mekong River is strongly influenced by the monsoon season. Due to the uneven seasonal distribution of rainfall, each year the basin experiences a differing intensity and duration of drought. This section investigates the water security of the riparian states that rely on the Mekong River.

China and its downstream neighbours have a great diversity in their levels of water security (see Table 5.2). China's total runoff area of the Mekong River reaches 26.1 per cent, the largest area of any of the countries of the Mekong River basin. However, the total withdrawal by China is only 10 per cent of the total withdrawal from the river basin's runoff area. Taking water quantity as a key criteria for examination, Yan (forthcoming) studied the level of water security of the Mekong River basin and found it to be relatively high. On the Mekong River, only Vietnam and Thailand have shown a water resources exploitation rate of over 10 per cent. The other four riparian states indicate a ratio of under 4 per cent, and China's ratio is kept at 3.7 per cent.

The LMCs have differing levels of reliance on the Mekong River. Each country has its own interests and focuses in water management, depending on its natural endowment. Yan (forthcoming) examined the river's water use from a sectoral perspective, by distinguishing between agriculture, industry and domestic water usage. Thailand's total withdrawal from the Mekong River reaches the highest level among the riparian states. It focuses on agricultural cultivation and exports, along with the growing demand for energy generation. Its water withdrawal in all three sectors ranks the highest among the LMCs. Vietnam ranks the second highest in its water use, with a large proportion of water resources consumed in irrigation. Myanmar has been less well connected with the rest of the LMCs but has shown a closer link with China. Proportionally, its industrial water withdrawal accounts for a larger percentage than that of the other riparian states. Laos sees a larger proportion of water consumption in the domestic sector, reflecting that it has been challenged to respond to its growing population. The country has been keen to

*Table 5.2* China and LMCs' water resource management in the Mekong River basin

| | | *River catchment area ($km^2$)* | *Runoff depth (mm)* | *Water resources ($10^9 m^3$)* | *Agricultural water withdrawal ($10^9 m^3$)* | *Industrial water withdrawal ($10^9 m^3$)* | *Domestic water withdrawal ($10^9 m^3$)* | *Total withdrawal ($10^9 m^3$)* | *Water resource utilization rate %* |
|---|---|---|---|---|---|---|---|---|---|
| Lancang River | Myanmar | 21,905.84 | 714.19 | 156.4500 | 2.0314 | 1.0428 | 0.1825 | 3.2567 | 2.1 |
| | Cambodia | 154,721.73 | 958.63 | 1,483.2078 | 26.0758 | 1.2558 | 0.4253 | 27.759 | 1.9 |
| | China | 16,474.62 | 454.89 | 749.4228 | 20.5108 | 3.7354 | 3.7975 | 28.0436 | 3.7 |
| | Laos | 206,248.73 | 1,339.72 | 2,763.1471 | 27.1569 | 2.5594 | 3.1379 | 32.8541 | 1.2 |
| | Thailand | 187,371.31 | 702.59 | 1,316.4506 | 129.3652 | 8.4962 | 13.1486 | 151.0100 | 11.5 |
| | Vietnam | 38,025.25 | 1,210.91 | 460.4516 | 62.1842 | 2.0260 | 4.4057 | 68.6158 | 14.9 |

Source: Yan (forthcoming).

develop hydroelectricity on the Mekong River to reduce its energy imports from other nations. Cambodia is rich in fish and aquatic species, hence it is less interested in developing damming projects. In contrast, Vietnam is interested in developing dams, but in order to protect its agriculture and aquaculture in the Mekong delta, it has shown a preference to dam on the distributary of the Lower Mekong River. Cambodia has also shown a high reliance on the river basin, with more than 70 per cent of the Cambodian population's annual animal protein consumption coming from the river's fish. Its agricultural and domestic water use has also reached a large proportion, compared with the river's other riparian states.

The Mekong River's riparian countries have also displayed differing levels of economic growth. Measured by per capita GDP, apart from China, Thailand is the wealthiest country on the river. Cambodia is the poorest while Laos, Vietnam and Thailand are in the middle of the range (see Table 5.3). With regard to the living standards and economic growth, some of the riparian states such as Myanmar and Vietnam represent the average national standards.

Concerns have been raised that China's dam construction on the Mekong River may be responsible for aggravating flooding or drought conditions (Hirsch, 2011). Studies indicate that the major causes of flood hazard are likely related to monsoonal rains or tropical cyclones (MRC, 2005; Ribbe *et al.*, 2013). According to these reports, there have been fewer man-made floods such as those related to dam construction than there have natural events such as monsoonal rains or tropical cyclones (MRC, 2005; Ribbe *et al.*, 2013). Damming activities are also suggested to have varying impacts on different downstream states. While discharge from dams may generate floods on the river's main stream that runs through Cambodia and Vietnam, it contributes significantly less further downstream. For instance, due to their geographical locations on the river, countries on the main stream are likely to experience more risks and hazards than other countries situated on the tributaries or local catchments (Joy, 2007). Laos, for instance, is found to have the sector most vulnerable to floods, with its paddy rice production being most threatened for over 20 days (Asian Disaster Preparedness Center, 2010; Ribbe *et al.*, 2013). Differing levels of vulnerability have also affected how these states have responded to China's damming

*Table 5.3* GDP per capita of communities on the Mekong River

| *Country* | *GDP per capita on the river (US$)* | *National GDP per capita (US$)* |
|---|---|---|
| Cambodia | 780 | 1,021 |
| Laos | 1,129 | 1,531 |
| Myanmar | 1,352 | 1,209 |
| Vietnam | 1,678 | 1,685 |
| Thailand | 3,777 | 5,775 |
| China | 4,336 | 6,498 |

projects. For instance, in 2011, rainfall throughout the flood season was the main cause for the flooding, and reports have shown that through its emergency control, China managed to keep the runoff level below an average level during the flood season (MRC, 2011).

It is also feared that damming projects have resulted in environmental impacts, with water quality and quantity being key concerns. Dam construction yields a poor prognosis for aquatic and floodplain health and it is the primary cause of degradation of freshwater ecosystems. Dams, together with the water reservoirs of Thailand, Laos and Cambodia, are believed to slow down the Mekong River's natural flow, and disrupt the existing equilibrium channel form or undercutting of banks. The downstream channel may scour, thereby become unstable. It is argued that the environmental impact of large-scale dams is destructive to the ecology of the river in downstream countries (Mehtonen *et al.*, 2008), where aquatic environment is destroyed (Kondolf *et al.*, 2014). Dams are likely to trap river sediments that harm the productivity of floodplain farms (Kummu *et al.*, 2010; Grumbine and Xu, 2011). This also affects the water's nutrient levels, the breeding cycles of fish, and the livelihoods of the coastal people. According to estimates, if all of the dams proposed for the river are built, there will be a 96 per cent loss in sediment load in the Mekong delta (Kondolf *et al.*, 2014), with serious impacts on downstream fisheries and food production systems. It has also been pointed out that damming on the Mekong River has the potential to impact on fisheries and other livelihood systems of downstream users of the river (Hirsch, 2011). Biodiversity and ecosystem functions across the Mekong River basin are under growing pressure and estuary areas have been particularly impacted (Hirsch, 2014).

## China's water diplomacy with the Lower Mekong Countries

### *China's positions and principles*

China regards the Mekong River as an essential resource for the development of the country's West, especially Yunnan Province. According to China's Twelfth Five-Year Plan, constructing a Mekong River port for international trade is one of the key targets for modernizing the Yunnan service sector and the development of the West (Zhang, 2015). In May 2011, China's State Council issued a document entitled 'Supporting the Accelerated Construction of Yunnan as the Important Outpost for the South-West Region'. It tasks the capital city of Kunming to become *the* international hub for China's south-western region facing the GMS.

In its foreign policy, China's position regarding the Mekong River has evolved significantly in recent years. In early times, its perception of the security of river basins differed greatly from that of its neighbours. By focusing on the Mekong River's quantity of water flows, it assumes the country to be relatively secured over this river, hence water-centred cooperation is not a primary concern in its foreign policy agenda.[3] However, it has been suggested that the

Beijing government aims to develop good-neighbourly diplomacy (Shi, 2013; Johnson *et al.*, 2014; Ruan, 2014), which is characterized by the primary goal of maintaining a peaceful peripheral environment for economic growth and political stability. China has shown an intention to adapt to concerns expressed by these countries.

The Chinese authorities have shown their willingness to comply to environmental laws in China's dam construction. With mounting discontent illustrated by environmentalists and international actors beyond China's borders, the Chinese government has made sporadic compromises in its decision-making for the Mekong River (see Box 5.1). For example, in one case it did halt dam construction due to fervent opposition raised by transnational environmental networks (Xie, 2009). Around 2003 when the Nu River damming project was advocated, a transnational environmental network was developed that extends to local environmental activists and scholars in Yunnan and beyond Chinese borders in some of the LMCs. The campaigns challenged that the policymaking process was not transparent and that there were grave concerns for community welfare in the LMCs. In fact, this project has become one of the few that has been permanently halted by the Chinese government. Part of the Nu River where damming was planned has now been preserved as a national park.[4] Nevertheless, China's decision-making process for dam construction remains non-transparent. It is uncertain what scientific evidence will be adopted by the Chinese authorities. The Chinese government is also not ready to incorporate multilevel basin management that will involve grassroots stakeholders in its decision-making for dam construction projects on the upstream of the Mekong River.

Along with its strategic interests constantly changing, China has revised its foreign policy agenda to suit its strategic interests in the region. China emphasizes voluntarism and consensus building in regional cooperation (Zhao, 2011). A regionalism approach is adopted to mobilize consensus among the LMCs. In particular, environmental issues, also referred as 'non-traditional security' issue, have been identified by the Chinese authorities as a common threat to the Asian countries. As the then premier Jiabao Wen claimed that China is 'deeply concerned over the increasing nature of non-traditional security issues … which have become important factors of uncertainty affecting regional and international security and are posing new challenges to regional and international peace and stability'.[6] The current president, Jinping Xi, has also reinforced this position, reassuring that the country is committed to developing close collaboration with the ASEAN countries on a series of non-traditional security issues, including climate change among other transnational issues.[7]

Furthermore, a notion of 'regionness' as well as a 'we-feeling' is important for mobilizing identity (Elliott, 2012). By stressing the nature of transnational environmental issues, the Chinese authorities highlight its commitment to peacemaking and regional cooperation in South-East Asia. With the wordings such as 'shared river, shared future',[8] the transboundary water issue is highlighted as a transnational issue of unexpected impacts across borders, upon

**Box 5.1 Disputes over water sharing: the 2011 Myitsone Dam project**

The Irrawaddy River is shared between China and Myanmar. The river basin covers nearly 60 per cent of Myanmar's land area. As the largest river in Myanmar, it is fed by three major rivers: the Chindwin, Shweli and Taping Rivers. These rivers were deemed suitable for a series of dams, to be built on the main stream of the Irrawaddy and its tributaries. Local residents on the downstream of the Irrawaddy largely rely on rice production and the river's fishery resources.

The Myitsone Dam was to stand at the headwaters of the Irrawaddy River. This is a US$8 billion project with a planned hydroelectricity generation capacity of 6,000 MW. It includes a series of seven dams that will reach a total cost of ¥200 billion and an overall hydroelectricity generation capacity of 20 million GW. The scale of the Myitsone Dam made it the largest hydropower project in Myanmar. The China Power Investment Corporation was responsible for its operation. According to a BOT (built-operation and transfer) mode of agreement, after 50 years of operation, the Myanmar government was to be transferred ownership of the dam that will generate energy for export to China.

Opposition to the dam was voiced once news of its construction was disclosed in 2007. The main local opposition came from environmentalists, who feared that the project had the potential to severely alter the downstream flows critical for rice production, aquatic migratory patterns and river salinity. The Kachin state government also showed reluctance to relocate, despite that financial compensation was offered. The Central Myanmar government was therefore confronted by opposition from various political forces such as local residents, national NGOs, the Kachin state government, the National League for Democracy party and ethnic groups which are organized by the Kachin Independence Army.

Sinohydro had commissioned an Environmental Impact Assessment (EIA) to be conducted by two independent research institutes, consisting of a team of 12 experts in fields including ecology, fisheries, environmental and social impact assessment, public health, flood management and hydrology.[5] A report was published online noting the potential impacts of the damming and public opinion towards the project. It found that the Myitsone Dam would destroy 1,200 ha of forests, arable land and other natural resources. The report also suggests an alternative plan for the construction of a series of smaller-sized dams that could replace the Myitsone Dam (Upstream Ayeyawady Confluence Basin Hydropower Co., 2011).

The disclosure of the EIA has further strengthened the opposition from the Kachin state government. Ethnic tensions in the region have posed a threat to the stability of the regime. It has not only contacted Hu Jintao, the then President of China, but has also escalated political tensions with the Chinese central government. The Kachin state government reiterated its rejection of the dam, leading to months of tension and violence between opposing parties within Myanmar.

The projects put Myanmar's then national government under much political strain. With President Thein Sein's military rule being challenged, fear was raised that instability would further escalate within the state of Kachin and the rest of the country (Li *et al.*, 2016). Thein Sein therefore halted construction of the dam. Nevertheless, rumours are that the Chinese government intends to restart or renegotiate the dam's construction with Myanmar's new leader, Aung San Suu Kyi.

which regional cooperation is required among riparian neighbours. The Chinese authorities have therefore advocated that the government commits to cooperating with the LMCs and to 'building a bond of cooperation and a common homeland' (Wen, 2008). By so doing, the Chinese government aims to show its intention to recognize the shared river as a key issue that the country needs to address with its riparian neighbours, and that it has adopted a peaceful depoliticizing position on the water-sharing issue.

### *China's involvement in the Mekong River Commission*

Following the dynamic changes in China's positions with the Mekong River riparian states, its diplomatic initiatives have changed. This is reflected in its interaction with regional cooperation mechanisms for basin-wide river management, with the two most important being the Mekong River Commission (MRC) and the Greater Mekong Subregion Mechanism (GMSM), an economic cooperation programme (see Box 5.2).

China's institutional cooperation with the MRC had been rather limited (see Table 5.1). Since 1992, China has developed formal diplomatic relations for dialogue with the MRC, but without becoming a member. China disagrees with

**Box 5.2 Water management institutions (WMIs) in the MRC**

The MRC was established to coordinate basin-wide management of the Mekong River. It was transformed by the Mekong Committee, which was a result of the USA's Mekong basin water policy in the 1960s. Since then, the Mekong riparian states have seen various donors involved in promoting basin-wide management initiatives made at an intergovernmental level. They are EU, European countries and Japan. Only four LMCs signed the Agreement on the Cooperation for the Sustainable Development of the Mekong River Basin on 5 April 1995 in Chiang Rai, Thailand. China and Myanmar have been Dialogue Partners of the MRC since 1996 and regularly attend the annual Dialogue Meetings and technical meetings.

The MRC consists of a Council at the ministerial level, a Joint Committee at the senior officials' level, and a Secretariat which serves as the technical and administrative arm of both the Council and the Joint Committee and under the direct supervision of the Joint Committee. Member countries have also established National Mekong Committees and secretariats to further support the MRC in its mission at the individual country level. In particular, it is highlighted that the donor community has a strong impact on the operation and decision-making within the MRC (Mehtonen *et al.*, 2008; Hensengerth, 2009b; Mirumachi, 2015). Apart from the newly elected chief executive officer of the MRC Secretariat, key personnel come from donor countries or agencies rather than a riparian nation, reflecting the donor-driven nature of the organization. The sponsorship feature of the MRC has also proved an obstacle in the MRC's operations (Sunchindah, 2005).

MRC's principle of consulting with LMCs for the management of water resources within its territory. China insists that it has complied with the principles adopted by the MRC that include 'developed equal relations, consultation, cooperation and common development', and the country aims to promote trust with the MRC and individual LMCs.[9] China's primary concerns lie in the fact that if it joined the MRC, it would lose the entitlements that it has long enjoyed as an upstream country (Lu, 2012). Accordingly, China has developed limited institutional diplomacy with the MRC. In recent years, along with its shifting of strategic interests, the country has shown a stronger commitment to information sharing. To reduce mutual suspicion and tension, China is now willing to share more data relating to the quantity and quality of water, particularly during the dry season. In 2002, an agreement was signed for China to share hydrological information on the Mekong River in the flood season.[10] This agreement promoted trust-building between China and the LMCs, with information-sharing being a key cooperative initiative between China and the MRC. However, hydrological data has been classified as restricted information and hence made unavailable to downstream countries. It was not until 2010 when actual disclosure of information began to occur. This came in the wake of disquiet in downstream countries over the possible impacts of reservoir-filling on river-flow reductions and of releases on flash floods (Hirsch, 2011).

### *China's involvement in the Greater Mekong Subregion Mechanism (GMSM)*

Compared with China's cooperation with the MRC, its engagement has strengthened with the GMSM. Joined by six countries, the GMSM is financially assisted by the Asian Development Bank, with the aim of strengthening economic linkages between the Mekong's riparian states. The GMSM's principle is 'mutual respect, equal consultation, practical results, gradual progress'.[11] Its schemes intend to bring benefit in terms of raising the income and productivity of the member countries. The development of water structures has been split into distinctive areas, such as hydropower, flood control and irrigation, and stream channelization for navigation purposes (Hensengerth, 2009a; Mirumachi, 2015).

The GMSM's intention to promote regional economic growth matches China's agenda to secure economic interests in the Mekong River basin. For instance, China's water diplomacy had oriented towards maximizing benefits from the sharing of the Mekong, hence it has shared with the LMCs' common interests in the subregional cooperation that are supported by a 'Mekong Spirit' that prioritizes economic connections in the GMSM. Within the GMSM, China is recognized to play an increasingly active role. The Beijing government has strategically promoted a common identity among riparian countries. For instance, at the Twelfth GMSM Ministerial conference and later in 2005 at the Second GMS Summit in Kunming, China strongly advocated regional economic integration programmes, for instance, through the construction of a

Mekong power grid. According to the Beijing government, the Mekong power grid represents one of the flagship initiatives of the Greater Mekong Subregion programme. The programme suits the LMCs' interests to prioritize obtaining profits from water policies (Hirsch and Jensen, 2006; Reilly, 2013). It aims to provide a regional grid as well as establishing a system for regional power trade through encouraging private investment in the power sector.[12] It is envisioned that China, Myanmar and Laos provide hydropower, a plan that has gained financial support from the Asian Develop Bank.

Nevertheless, the Chinese government realizes that its position in the GMSM remains awkward. On the one hand, China eases financial investment as well as policy learning. Its knowledge and institutional capabilities have facilitated such a process, when, for instance, it transferred its environmental policies and legal framework to neighbouring riparian states.[13] On the other hand, the Chinese government sees it best to retain a non-member status in order to pursue its national interests. The Beijing government regards the Asian Development Bank as a competitor that constrains it from negotiating smoothly with the GMSM.[14] Additionally, China's involvement in implementing a cooperative programme within the GMSM is also constrained by complex politics in riparian states, including corruption at the local government level.[15] China has complied with the approach taken by members of the GMSM, which is based on a 'soft and ambiguous' agreement and shows no attempt to coordinate the conflicting interests (Hirsch and Jensen, 2006). For instance, among its members, there lacks coordinated efforts to collectively regulate the management of the Mekong's water resources (Hensengerth, 2009a; Mirumachi, 2015). With a limited role to play, China was bound to develop new forms of diplomatic cooperation with the LMCs.

### *Regional economic integration and China's strategic interests over the Mekong River*

With China's increasing commitment to cooperation within the GMSM, it has become proactive in managing its transboundary interests. Such moves are vested with political and strategic interests. The country's demand for regional stability and leadership has fuelled its interest in taking on a leading role among the Mekong River's riparian actors. More recently, it has established the Lancang-Mekong Cooperation Mechanism (LMCM), within which the country has assumed a greater role. This mechanism shows a similarity to pre-existing mechanisms. For instance, it has continued to support the position whereby the state is the dominant actor in basin-wide mechanisms (Gain and Schwab, 2012; Suhardiman and Giordano, 2014).

With the LMCM, China has brought forward its economic interests on the Mekong River. It is also evident that, with this mechanism, China has engaged and negotiated with the LMCs, and on some occasions agreements have been made to collectively manage the Mekong River (see Table 5.1). With regard to the content, hydropower development and navigation are prioritized in the scheme. As China's positions with its fellow riparian actors evolve, it has gradually expanded its agenda to tackle basin-wide water management issues. Individual issues in water management have also been incorporated to reflect common interests in regional stability and economic development, among which disaster control and biodiversity protection are two identified areas.

### *Dam construction*

The LMCM facilitates China to bring on board the Mekong's other riparian nations. Dam construction is a key issue for the Beijing government. Since the 1990s, Chinese investment has grown in terms of the total investment in hydropower development in the Lower Mekong River. To date, six dams on the Mekong River's main stream have been built (see Table 5.4). According to the government's Thirteenth Five-Year Plan, more than 2,500 MW is to be harnessed from the Mekong River by 2020. China's increasing scale of hydropower investment has been accompanied by the country's development of associated infrastructure, including a regional power grid.

Although there has been opposition to China's dam construction projects, the responses of individual LMCs have varied. To some degree, the LMCs welcome funding for public investment from China, which differs in its requisitions from other funders such as the World Bank and the Asian Development Bank (Hirsch, 2011). With China, the LMCs are found to succeed in prioritizing economic imperatives in their planning and management of the Mekong River basin (see also Fox and Sneddon, 2007). As little coordination has been achieved via existing basin-wide management institutions, with the interest of financial gains, China has managed to secure consent from its neighbouring riparian states for most of its projects (Suhardiman *et al.*, 2012). In particular, in Cambodia, China acts as the main financier for most of the dams built, with state-owned enterprises dominating all projects (Urban *et al.*, 2015).

### *Navigation*

China has been involved in promoting cross-border trade and transport with the LMCs. Enhancing waterway safety is another main area of focus.[16] With respect

*Table 5.4* Dams on the Mekong

| *Dam* | *Output (MW)* | *Height (metres)* | *Year completed* |
|---|---|---|---|
| Manwan | 1,550 | 126 | 1993 |
| Dachaoshan | 1,350 | 118 | 2003 |
| Jinghong | 1,650 | 107 | 2009 |
| Gongguoqiao | 750 | 105 | 2012 |
| Xiaowan | 4,200 | 292 | 2012 |
| Nuozhadu | 5,850 | 262 | 2014 |

to waterway navigation, citizen safety has become a key focus in China's cooperation with the LMCs. As early as 1993, China signed the Quadripartite Economic Cooperation agreement. This agreement has resulted in the most substantial changes to the downstream parts of the Mekong. With Laos, Burma and Thailand, China has agreed to jointly survey the Mekong River basin to ensure that any navigation project is well-designed and monitored. The thrust of these collaborative programmes has so far predominantly focused on promoting and facilitating economic and infrastructure development.

China has initiated a number of collaborative projects with individual LMCs. With Laos, work was carried out to clear rapids, shoals and reefs to canalize the basin and allow smooth navigation from Simao (China) to Luang Prabang (Laos) (IRN, 2002; Mirumachi and Nakayama, 2007). Following a major programme to clear obstacles from the Mekong that began early in the 2010s, a regular navigation service now exists between southern Yunnan and the northern Thai river port of Chiang Saen. Similarly, with Myanmar and Thailand, China has also facilitated the construction and renovations of navigation channels and other measures to eliminate navigational dangers.[17] A major component of this project has been the building a 71 km five-star navigation channel along the Mekong River from Jinghong to the Myanmar border. This project has also resulted in the completion of a channel, navigable by 300-tonne vessels, that links China with all of the LMCs. China has further continued with initiatives to expand trade activities by improving the navigation channels, and this includes activities that may have social and environmental impacts. It has proactively promoted the signing of agreements with another three countries, Myanmar, Laos and Thailand. However, there were concerns that such projects had not taken into consideration the livelihood and ecological impacts along the river (IRN, 2002).

China has also shown a strong commitment to work with chosen LMCs to address smuggling and the safety of transnational navigation. Since the 2011 Mekong Massacre,[18] the country has been more alert to transnational security issues on the Mekong River. Approximately 90 per cent of Asia's second largest opium-producing route runs along Laos' borders with Myanmar and Thailand. This region is known as the Golden Triangle. The drug smugglers and dealers in this area have long posed a threat to international shipping on the waterway, especially in the north-eastern areas of Myanmar that are under the control of ethnic armies.

China's proposed diplomatic cooperation with Laos, Myanmar and Thailand includes information sharing to improve communications between the countries and the establishment of a new police unit, the Waters Division, to enforce control. With safety on the transnational river now better controlled, China's cooperative initiative has won trust from the collaborating countries. To some extent, this water diplomacy has promoted intra-issue links in China's foreign policies to address the arising concerns in international affairs on the Mekong River.

*Disaster control*

To reduce the mistrust of the LMCs, China signed a data-sharing agreement with the MRC in 2002 and then renewed it in 2008, 2013 and 2016. Since then, China has provided precipitation and water-level data collected from its two hydro-meteorological stations located on the Mekong's main stream at Yunjinghong and on a tributary at Manan. Since China has become more aware of the importance of information-sharing for disaster prevention in the LMCs, it has agreed to extend the dates of hydrological data provision and increase the frequency of the information exchange.

China has also promptly responded to water management emergencies in the LMCs, although such cooperation has been and is likely to continue to be on an ad hoc basis. For instance, China has frequently provided emergency water supplies to alleviate drought. In 2016 Vietnam's Bin Tre Province in the southern Mekong delta was seriously was hit by drought. China responded to Vietnam's calls for aid by discharging from Jinghong twice per day the amount of water normally discharged from the Jinghong Hydropower Station. The volume of water discharged from 15 March to 10 April was three to three-and-a-half times more than the normal flow.[19]

Another form of cooperation is for China to share its disaster management experiences and policies, including in relation to droughts and flooding. For example, the Ministry of Water Resources has provided training and workshops to share its know-how and management experiences on flood prevention with the MRC and professionals from individual LMCs.[20] Nevertheless, the extent that LMCs will benefit from techniques in the prevention of dam-related floods remains unknown, as these countries have shown a weak technological capacity to respond to water-related events, such as catastrophic monsoonal rains.

*Biodiversity protection*

China has shown an increasing commitment to biodiversity protection. It has become aware that biodiversity loss is one of the negative impacts of its dam construction projects on the main stream of the Mekong River (Zhong *et al.*, 2007). This could be related to the country's interest in boosting regional development in Yunnan and Guangxi. At the same time, the move has political implications and would promote China as a committed participant to the GMSM.[21] To protect the Indo-Burma biodiversity hotspot, as early as 2005 China joined a pilot programme coordinated by the GMSM called the Biodiversity Conservation Corridors Initiative (BCI). Under this initiative, China has signed agreements with Vietnam and Laos to establish nature reserves and implement other transboundary conservation measures such as biodiversity corridors. Its provincial governments in Guangxi and Yunnan have developed international partnerships to promote the establishment of integrated development and conservation activities. For example, Guangxi Province in China and Cao Bang Province in Vietnam have cooperated on the cross-border management of important habitats of the

Cao vit crested gibbon by establishing a biodiversity corridor called the Jingxi Transboundary BCI site (Kano *et al.*, 2016), and Yunnan has worked collaboratively with Laos in its border region to establish the Xishuangbanna BCI pilot site. After these pilot programmes had been completed, continued institutional support was provided by the Chinese government to establish a national support unit to coordinate cooperation and activities with the neighbouring partners. This indicates that substantive institutional cooperation has been developed on the protection of biodiversity. Overall, the programmes have focused not only on ecosystem restoration but also poverty reduction and land-use planning and management. This has resulted in village development funds and forest regeneration, as well as corridor boundary approval. Apart from the national and local governments, national NGOs and international actors have also been incorporated into the projects. However, it is noted that at the grassroots level, there is a lack of legal instruments and institutional arrangements, and the collaborative programmes' outcomes for the protection of biodiversity have therefore been rather slow and inefficient (Zhang and Zhao, 2011; Wang *et al.*, 2014).

### *Implementation activities that are facilitated by aid programmes*

China has adopted pragmatic measures to facilitate its foreign investment and trade. In the country's promotion of trade, foreign direct investment (FDI) and official development assistance (ODA) have been a featured approach, with the aim of quelling negative diplomatic and security disputes in South-East Asia. However, whether such an approach has succeeded remains debatable (Kalathil, 2012). Similarly, financial assistance is a strong focus in China's water diplomacy (Lu, 2016). From a realism perspective, China views the externally funded mechanisms to be dependent on funders' interests and becoming inadequate at implementing their own agendas that have been agreed by the member states. The Beijing government feels increasingly pressured by the competing initiatives in the region. It has been concerned that the existing basin management mechanism had enjoyed little independence, as donor influence over the institutional mechanisms was regarded as too strong,[22] such as in the cases of the Mekong–Ganga Initiative (established by India in 2000), the Japan–Mekong Regional Partnership (established by Japan in 2007), the Lower Mekong Initiative (established by the USA in 2009), and the Mekong–Korea Comprehensive Partnership for Mutual Prosperity (established by South Korea in 2011). These mechanisms are seen as competing with China for strategic interests in South-East Asia (Li, 2011). Hence, the Beijing authorities have expedited their funding of the LMCM to develop planning and management policies for the Mekong River.[23] Several forms of aid have been provided.

- Direct financial flows
  In China's relations with the ASEAN countries, providing financial assistance for poverty relief has been an essential aspect to strengthen peaceful relations. Similarly, providing large sums of available funding is deemed an

essential part of its water diplomacy,[24] particularly when China's planning and management over the Mekong requires a large financial investment. The Chinese government boasts that its aid programme has exceeded LMCs benefits from previous donors such as the Asian Development Bank or other US funders (Lu, 2016). It also argues that it provides a broader scale of aid programme, and that its implementation has been solid. To facilitate the launching of the LMCM, Premier Li Keqiang made a series of proposals and commitments concerning promoting cooperation.[25] The offer includes US$10 billion for infrastructure interconnectivity, US$490 million in grants for poverty alleviation, and US$1.6 billion in special loans for enhancing China's export production capacity. Such a financial investment gives the Chinese government more leeway to coordinate various interests (Zhu, 2008; Reilly, 2013; Urban *et al.*, 2013). Individual riparian countries such as those suffering from serious poverty have also received tailor-made treatments. For instance, the Chinese government has provided grant funding of over ₭34 billion (over ¥33 million or some US$4 million) to address poverty reduction in Laos, especially in outlying areas of the capital Vientiane and rural areas of Luang Prabang Province.[26]

- Infrastructure investment
  As China has developed close economic ties with most of the LMCs, it has exhibited a strong interest in boosting its investment in industrial development. The country is eager to provide funding to expedite the development of physical infrastructure in the GMSM member countries. Along the Red River and the Mekong, a water navigation and transport project was established, and a joint venture business to build dams has been developed between China and Myanmar. The infrastructure investment has also financed projects in which China has accepted a leading role in the management of the Mekong River's water resources.[27] For instance, China's initiative to prevent natural disasters along the Mekong River has often been bundled together with railway development that has benefited international trade and investment.[28]
- Technological capacity building
  Apart from profits gained in selling electricity to LMCs, China's involvement includes technological cooperation. This involves the transfer of skills, knowledge and expertise for innovation (Hirsch and Jensen, 2006; Urban *et al.*, 2015), which constitutes an important part in its implementation of its water policies over the Mekong River. Such activities are based on FDI or ODA that involve company cooperation in the form of mergers and acquisitions. Urban *et al.*'s research (2015) is one of the earliest studies to have investigated technology transfer and technological cooperation between China and other developing states. Chinese damming companies are willing to transfer knowledge of engineering, management and technological capabilities to their partner states. However, the outcomes of technology cooperation are far from ideal. Among various factors contributing to this, one of the largest is the recipient country displaying a weak capacity to absorb the new knowledge.

## Concluding remarks

The Mekong River, which is a significant contributor to the livelihoods and well-being of the communities that live along it, is faced with an escalating level of violence and conflicts. However, there is a lack of institutions to effectively coordinate the management of the river. In particular, China, located on the upstream, has been viewed as a hegemon that takes advantage of the downstream states. This is not only related to China's position as an upstream country with geographical advantage, but also to the political mistrust towards China as a regional power. This view is shared among China's downstream neighbours.

When evaluating China's water diplomacy, it is evident its practices have been successful in promoting regional cooperation, when China has engaged and negotiated with the LMCs, and on some occasions agreements have been made to collectively manage the Mekong River. The Chinese national government has been efficient in establishing a new mechanism to facilitate state-led cooperation, which combines trade, infrastructure development and other arising issues in the planning and management of water resources.

Scholarship indicates that pre-existing cooperation mechanisms play significant roles in promoting the formation of water management institutions (WMIs) (LeMarquand, 1977; Bernauer, 1997; Wolf *et al.*, 2003; Conca *et al.*, 2006). This is particularly the case with those pre-existing mechanisms that promote economic cooperation, which is deemed to be the most successful means of facilitating institutionalized cooperation (Dosch and Hensengerth, 2005; Tir and Ackerman, 2009). In this case, the findings have confirmed that in the context of the global South, it is likely that new institutional cooperation is formed, particularly when interests in regional development and economic growth have been incorporated. Nevertheless, despite that institutional cooperation has been formed among all countries that share the Mekong River, China and its involvement in the LMCM does not necessarily prioritize basin-wide water management in its national agenda.

The empirical report has shown that China has vested various interests in transboundary water-sharing that serves its economic and strategic interests. Regional development of Yunnan and Guangxi Provinces has been a priority in how it appropriates the Mekong River's water resources. Reflected in national policy, China has shown its strong intention to economically exploit the river's water resources. However, as the country is aware of the concern shown by its riparian neighbours, it has also made it apparent that maintaining peace is a key national interest in regional development. For this purpose, various approaches are adopted, including establishing regional cooperation through the multilateral mechanism of the LMCM. China also relies on the LMCM to balance regional opponents that include Japan and the USA. Hence, water-sharing is a politicized issue and the aim of the country's water diplomacy is multifaceted.

Unlike other forms of cooperation on regional affairs, on the issue of water cooperation, China's efforts may prove to be instrumental to establishing basin-wide management. China's water diplomacy largely depends on the country being an economic power that is capable of providing financial assistance, technical support and other material interests to its riparian neighbours. On a short-term basis, the riparian actors are incentivized to participate in the LMCM as their interests have been driven by seeking to simultaneously secure financial income and national economic development. Therefore, the LMCM agenda on basin-wide management is likely to facilitate collective actions on horizontal level.

Nevertheless, China, and the LMCM under its influence, is faced with multiple challenges. On the development of the Mekong River basin, the lack of organic and effective, vertically integrated institutions among these states (Suhardiman *et al.*, 2012; Hensengerth, 2015). With national policy process being fragmented there lacks effective governing arrangements to transform regional policy into domestic policy. This situation has become a source for both interstate and intra-state conflicts for the riparian states (Molle *et al.*, 2009; Suhardiman *et al.*, 2012). Furthermore, China faces the challenge, in its leading of the LMCM, to effectively incorporate the wide range of stakeholders that are involved in the development of the Mekong River basin. Apart from those located in the region and relevant riparian states, non-state actors constituted of international and local NGOs are another important actors. Issues exist in the governance of the Mekong River – for instance, articulation of water injustice arises, pertaining to unequal access to water and unequal distribution of risks and the negative impacts from large-scale infrastructure development. As the country has shown the intention to occasionally incorporate NGOs in project implementation, the LMCM may provide opportunities to incorporate non-state actors at both regional and international levels in order to promote sustainable and just development.

In the long run, uncertainties remain in China's water diplomacy with the LMCs. With these countries exhibiting varying levels of water security on the Mekong River, the LMCs' water diplomacy with China evolves rapidly, as each incorporates their own strategic and economic interests. From this perspective, it has become increasingly important for China to negotiate with the LMCs via the LMCM, if it remains in its leading role. The country's resources may be overstretched if it tries to accommodate those riparian neighbours that are weak in their management of water. Before that occurs, China is faced with challenges to best coordinate various riparian actors and to effectively promote efficient institutions to realize sustainable basin management on the Mekong River. Otherwise, China, among other states that share the Mekong River, is faced with the danger of the river basin being fraught with intra-state and interstate conflicts as a result of degraded water management.

## Notes

1 ASEAN is a coalition of ten countries in the South-East Asia region. These include Indonesia, Malaysia, the Philippines, Singapore and Thailand, and now also Brunei, Cambodia, Laos, Myanmar (formerly Burma) and Vietnam after the organization's membership was expanded.

2 According to this Agreement, China will participate in hydroelectric cooperation with Myanmar through implementation of the BOT Ruilijiang First Class Hydroelectric Power Station with an installed capacity of 600,000kW. However, this cooperation has stagnated since the government of Myanmar set aside the Myitsone Dam Project.
3 Personal communication with Department of Asia, Ministry of Foreign Affairs, 2 May 2016.
4 Zhao, M. 2016. Yunnan government approves 13 national natural conservation park, *The Paper*, 19 September. Available at: www.thepaper.cn/newsDetail_forward_1530425 (accessed 2 February 2017).
5 The International Rivers organization has documented the Environmental Impact Assessment for the Myitsone Dam, which is available at: www.internationalrivers.org/resources/environmental-impact-assessment-for-the-myitsone-dam-2449 (accessed 2 February 2017).
6 See full text of the Joint Declaration of ASEAN and China on cooperation in the field of non-traditional security issue, Sixth ASEAN–China Summit, Phnom Penh, 4 November 2002. Available at: www.vifindia.org/sites/default/files/Joint%20Declaration%20of%20ASEAN%20and%20China%20on%20Cooperation%20in%20the%20Field%20of%20Non-Traditional%20Security%20Issues,%202002.pdf (accessed 29 May 2017).
7 See full text of Chinese President Jinping Xi's speech at Boao Forum for Asia Annual Conference 2015, 'Towards a community of common destiny and a new future for Asia', 28 March 2015. Available at: http://news.xinhuanet.com/english/2015-03/29/c_134106145.htm (accessed 29 May 2017).
8 Speech given by Chinese premier Wen Jiabao's at the Third GMS Summit, Vientiane, Laos, 31 March 2008. Available at: http://wcm.fmprc.gov.cn/pub/chn/pds/ziliao/zt/dnzt/wmgh/t419795.htm (accessed 28 August 2017).
9 Chinese Vice Foreign Minister Song Tao's talk with the CEO of the Mekong River Commission Jeremy Bird, at the first Mekong River Commission (MRC) Summit in Hua Hin, Thailand. Available at: www.mrcmekong.org/news-and-events/speeches/first-mrc-summit-5/ (accessed 28 August 2017).
10 The agreement has been renewed in 2008.
11 See Chinese government's publication, *Country Report on China's Participation in Greater Mekong Subregion Cooperation*, 2008. Available at: http://en.ndrc.gov.cn/newsrelease/200804/t20080430_208063.html#_Toc193801979 (accessed 2 February 2017).
12 Report produced by International Rivers Network, *Sizing Up the Grid: How the Mekong Power Grid Compares Against the Policies of the Asian Development Bank*, January 2004. Available at: www.internationalrivers.org/sites/default/files/attached-files/sizing_up_the_grid_how_the_mekong_power_grid_compares_against_the_policies_of_the_asian_development_bank.pdf (accessed 29 May 2017).
13 Interview with Peng Ren, project manager from Global Environment Institute, August 2016.
14 Personal communication with Department of Asia, Ministry of Foreign Affairs, 2 May 2016.
15 Ibid.
16 See Mekong River Commission's introduction on its navigation programme. Available at: www.mrcmekong.org/about-mrc/programmes/navigation-programme/ (accessed 2 February 2017).
17 Ibid.
18 On the morning of 5 October 2011, two Chinese cargo ships were attacked on the Mekong River in the Golden Triangle region on the borders of Myanmar and Thailand. All 13 crew members on the two ships were killed and their bodies dumped into the river.

19 Emia, V., China sends water to help drought-stricken Vietnam, 8 April 2016, *Yibada*. Available at: http://en.yibada.com/articles/114758/20160408/china-sends-water-help-drought-stricken-vietnam.htm#ixzz4XFZ4W9ls (accessed 2 February 2017).
20 China increases technological cooperation with the Mekong River riparian states on water management, 25 November 2011, *China News*. Available at: www.chinanews.com/gn/2011/11-25/3487521.shtml (accessed 2 February 2017).
21 See Chinese government's publication, *Country Report on China's Participation in Greater Mekong Subregion Cooperation*, 2008. Available at: http://en.ndrc.gov.cn/newsrelease/200804/t20080430_208063.html#_Toc193801979 (accessed 2 February 2017).
22 Personal communication with Department of Asia, Ministry of Foreign Affairs, 2 May 2016.
23 Ibid.
24 Wang, Q., Six countries agree on Mekong cooperation system, 11 December 2015, *China Daily*. Available at: www.chinadaily.com.cn/world/2015-11/12/content_22443585.htm (accessed 2 February 2017).
25 East Asia Cooperation Initiative on Poverty Reduction proposed by Chinese Premier Li Keqiang at the Seventeenth ASEAN Plus Three Summit held in November 2014.
26 Chandara, P., China funds poverty reduction in Laos, *Vientiane Times*, 8 December 2016. Available at: www.vientianetimes.org.la/FreeContent/FreeConten_China286.htm (accessed 2 February 2017).
27 Cambodia welcomes Chinese premier's pledge of aid to GMS countries, *Xinhua News*, 20 December 2014. Available at: www.chinadaily.com.cn/world/2014livisitkst/2014-12/20/content_19133676.htm (accessed 2 February 2017).
28 According to Keqiang Li, US$16.4 million was to be invested to dredge waterways along the Mekong River.

## References


Asian Disaster Preparedness Center, Public Works and Transport Institute, Lao PDR and UNDP. 2010. *Developing a National Risk Profile of Lao PDR*, *GRIPWEB*, pp. 1–94.

Bernauer, T. 1997. Managing international rivers, in O. Young, ed., *Global Governance: Drawing Insights From the Environmental Experience*, Cambridge, MA: MIT Press, pp. 155–195.

Conca, K., Wu, F. and Mei, C. 2006. Global regime formation or institution building? The principled content of international river agreements, *International Studies Quarterly*, 50(2): 263–285.

Dosch, J. and Hensengerth, O. 2005. Sub-regional cooperation in Southeast Asia: The Mekong Basin, *European Journal of East Asian Studies*, 4(2): 263–286.

Elliott, L. 2012. ASEAN and environmental governance: Strategies of regionalism in Southeast Asia, *Global Environmental Politics*, 12(3): 38–57.

Fox, C. and Sneddon, C. 2007. Transboundary river basin agreements in the Mekong and Zambezi basins: Enhancing environmental security or securitising the environment, *International Environmental Agreements: Politics, Law and Economics*, 7(3): 237–261.

Gain, A. and Schwab, M. 2012. An assessment of water governance trends: The case of Bangladesh, *Water Policy*, 14(5): 821–840.

Goh, E. 2014. The modes of China's influence: Cases from Southeast Asia, *Asian Survey*, 54(5): 825–848.

Grumbine, R.E. and Xu, J. 2011. Mekong hydropower development, *Science*, 332(6026): 178–179.

Hensengerth, O. 2009a. *Regionalism in China–Vietnam Relations: Institution-Building in the Greater Mekong Subregion*, Abingdon: Routledge.

Hensengerth, O. 2009b. Transboundary river cooperation and the regional public good: The case of the Mekong River, *Contemporary Southeast Asia: A Journal of International and Strategic Affairs*, 31(2): 326–349.

Hensengerth, O. 2015. Where is the power? Transnational networks, authority and the dispute over the Xayaburi dam on the lower Mekong mainstream, *Water International*, 40(5): 911–928.

Hernandez, C.G. 2008. The ASEAN charter and implications for East Asian regionalism, *Panorama*, 2: 25–36.

Hirsch, P. 2011. China and the cascading geopolitics of lower Mekong dams, *The Asia-Pacific Journal*, 9(20): art. 2. Available at: http://apjjf.org/2011/9/20/Philip-Hirsch/3529/article.html (accessed 28 August 2017).

Hirsch, P. 2014. The politics of engaged geography on the Mekong, *ACME: An International Journal for Critical Geographies*, 13(4): 516–524.

Hirsch, P. and Jensen, K. 2006. *National Interests and Transboundary Water Governance of the Mekong*, Australian Mekong Resource Centre in association with DANIDA and the University of Sydney, Australia.

IRN (International Rivers Network). 2002. *Navigation Project Threatens Livelihoods, Ecosystem*, IRN Briefing Paper No. 2, Berkeley, CA.

Johnson, C., Bower, E., Cha, V., Green, M. and Goodman, M. 2014. *Decoding China's Emerging 'Great Power' Strategy in Asia*, Washington, DC: Center for Strategic and International Studies.

Joy, C. 2007. Best practice IFRM guidelines for the basin development program. MRC flood management and mitigation program. In MRC (2010) *State of the Basin Report 2010*. Phnom Penh: Mekong River Commission.

Kalathil, S. 2012. Influence for sale? China's trade, investment and assistance policies in Southeast Asia, *East and South China Seas Bulletin* 4, Washington, DC: Center for a New American Security.

Kano, Y., Dudgeon, D., Nam, S., Samejima, H., Watanabe, K., Grudpan, C., Grudpan, J., Magtoon, W., Musikasinthorn, P., Nguyen, P.T., Praxaysonbath, B., Sato, T., Shibukawa, K., Shimatani, Y., Suvarnaraksha, A., Tanaka, W., Thach, P., Tran, D.D., Yamashita, T. and Utsugi, K. 2016. Impacts of dams and global warming on fish biodiversity in the Indo-Burma hotspot, *PLOS One*, 11(8): e0160151.

Kondolf, G.M., Rubin, Z.K. and Minear, J.T. 2014. Dams on the Mekong: Cumulative sediment starvation, *Water Resources Research*, 506(6): 5158–5169.

Kummu, M., Lu, X., Wang, J. and Varis, O. 2010. Basin-wide sediment trapping efficiency of emerging reservoirs along the Mekong, *Geomorphology*, 119(3): 181–197.

LeMarquand, D.G. 1977. *International Rivers: The Politics of Cooperation*, Vancouver: Westway Research Centre, University of British Columbia.

Li, C., Sein, C.C. and Zhu, X. 2016. *Myanmar: Reintegrating Into the International Community*, Singapore: World Scientific.

Li, Z. 2011. Balancing U.S.–Japanese Power in the Mekong River Basin, *China and World Affairs*, 23: 149–158.

Liebman, A. 2005. Trickle-down hegemony? China's 'peaceful rise' and dam building on the Mekong, *Contemporary Southeast Asia: A Journal of International and Strategic Affairs*, 27(2): 281–304.

Lu, G. 2012. Advantages and disadvantages in China's joining the MRC [Zhongguo Jiaru MeigongheWeiyuanhui Libi ruhe], *World Affairs*, 8: 30–32.

Lu, G. 2016. Lancang-Mekong cooperation: How did it emerge from all the multilateral mechanisms? *Contemporary World*, 3: 20–23.

Mehtonen, K., Keskinen, M. and Varis, O. 2008. The Mekong: IWRM and institutions, in O. Varis, C. Tortajada and A.K. Biswas, eds, *Management of Transboundary Rivers and Lakes*, Berlin: Springer, pp. 207–226.

Mirumachi, N. 2015. *Transboundary Water Politics in the Developing World*, Abingdon: Routledge.

Mirumachi, N. and Nakayama, M. 2007. Improving methodologies for transboundary impact assessment in transboundary watercourses: Navigation channel improvement project of the Lancang-Mekong river from China–Myanmar boundary marker 243 to Houei Sai of Laos, *International Journal of Water Resources Development*, 23(3): 411–425.

Molle, F., Lebel, L. and Foran, T. 2009. Contested Mekong waterscapes: Where to next? In F. Molle, T. Foran, and M. Käkönen, eds, *Contested Waterscapes in the Mekong Region: Hydropower, Livelihoods and Governance*, Abingdon: Earthscan, pp. 383–405.

MRC (Mekong River Commission). 2005. *Overview of the Hydrology of the Mekong Basin*, Vientiane: Mekong River Commission.

MRC (Mekong River Commission). 2011. *Flood Situation Report 2011*, MRC Technical Paper No. 36, November, Phnom Penh: Mekong River Commission.

Reilly, J. 2013. *China's Economic Statecraft: Turning Wealth Into Power*, Lowy Institute for International Policy Analysis. Sydney: Lowy Institute for International Policy. Available at: http://sydney.edu.au/arts/government_international_relations/downloads/documents/Reilly_China_Economic_Statecraft.pdf (accessed 28 August 2017).

Ribbe, L., Nauditt, A., Meinardi, D., Morbach, M. and Becker, R. 2013. *Comparison of Key Drivers Regarding Their Significance for Hydro-Meteorological Extremes and Their Impacts on Selected Hotspots Within the Mekong River Basin*, Cologne: Institute for Technology and Resources Management in the Tropics and Subtropics, Cologne University of Applied Sciences. Available at: www.giz.de/en/downloads/giz2013-en-hydro-meteorological-extremes-mekong.pdf (accessed 21 August 2017).

Ruan, Z. 2014. Zhongguo xuyao goujian zengyang de zhoubian, *Guoji Wenti Yanjiu*, 2: 11–26.

Shi, Y. 2013. Zhongguo zhoubian xingwei zhong cengyoude 'shengli zhuyi': Dongneng he juece fuzaxing, *Xiandai Guoji Guanxi*, 10: 3–5.

Suhardiman, D. and Giordano, M. 2014. Legal plurality: An analysis of power interplay in Mekong hydropower, *Annals of the Association of American Geographers*, 104(5): 973–988.

Suhardiman, D., Giordano, M. and Molle, F. 2012. Scalar disconnect: The logic of transboundary water governance in the Mekong, *Society and Natural Resources*, 25(6): 572–586.

Sunchindah, A. 2005. Water diplomacy in the Lancang–Mekong River Basin: Prospects and challenges, Workshop on the Growing Integration of Greater Mekong Sub-regional ASEAN States in Asian Region, 20–21 September, Yangon, Myanmar.

Tir, J. and Ackerman, J.T. 2009. Politics of formalized river cooperation, *Journal of Peace Research*, 46(5): 623–640.

Upstream Ayeyawady Confluence Basin Hydropower Co. 2011. Publication of environmental impact report of hydropower development in upper reaches of Ayeyawady river. Available at: www.uachc.com/Liems/esite/content/showDetail.jsp?nid=6897&newtype_no=2248 (accessed 29 May 2017).

Urban, F., Nordensvärd, J., Khatri, D. and Wang, Y. 2013. An analysis of China's investment in the hydropower sector in the Greater Mekong Sub-region, *Environment, Development and Sustainability*, 15(2): 301–324.

Urban, F., Siciliano, G., Sour, K., Lonn, P, Tan, M. and Mang, G. 2015. South–South technology transfer of low-carbon innovation: Large Chinese hydropower dams in Cambodia. *Sustainable Development*, 23(4): 232–244.

Wang, W., Tian, Y., Chang, M. and Li, J., 2014. A review of transboundary protected areas network establishment, *Acta Ecologica Sinica*, 34(6): 1391–1400.

Wen, J. 2008. Build a bond of cooperation and a common homeland. Speech at the Third GMS Summit, Vientiane, Laos, 31 March. Available at: www.china-embassy.org/eng/gyzg/t458515.htm.

Wolf, A.T., Yoffe, S.B. and Giordano, M. 2003. International waters: Identifying basins at risk, *Water Policy*, 5(1): 29–60.

Xie, L. 2009. *Environmental Activism in China*, Abingdon: Routledge.

Yan, J. Forthcoming. Assessment of China's transboundary water resources and water use, doctoral thesis, Chinese Academy of Sciences, Beijing.

Zhang, C. 2015. An analysis development of modern service sector and a review of factors that affects its growth [Yunnan sheng xiandai fuwuye de fazhan xianzhuang jiqi yingxiang yinsu yanjiu], master's thesis, Yunnan University of Finance and Economics.

Zhang, L., Lu, G. and I. Baird. 2016. Crisis and trust building in the cooperation between China and Lancang-Mekong River [Zhongguo zai lancangjiang Meigonghe kuajie shui ziyuan hezuo zhongde xinren weiji yu huxinjianshe], *Indian Ocean Economic and Political Review*, 2: 16–27.

Zhang, S. and Zhao, X. 2011. GMS biodiversity conservation corridor legal issues [Dameigonghe ciquyu shengwuduoyangxing baohu zoulang jianshe falv wenti yanjiu], *Ecological Economy*, 7: 151–153 [in Chinese].

Zhao, S. 2011. China's approaches toward regional cooperation in East Asia: Motivations and calculations, *Journal of Contemporary China*, 20(68): 53–67.

Zhong, H., Liu, H. and Geng, L. 2007. The ecological impact of multiple dam construction on the Mekong River [Lancang liuyu tiji kaifa de shengtai huanjing leiji xiaoying], *Journal of Hydraulic Engineering*, S1: 577–581.

Zhu, Z. 2008. China's economic aid to CLMV and its economic cooperation with them, in M. Kagami, ed., *Report of the Study Group on Japan and China Economic Relations With Mekong River Basin (MRB) Countries*, BRC Research Report, No. 3. Bangkok: Bangkok Research Centre, IDE-JETRO.

# 6 South Asia

## China's evolving attitude over the GBM

### Introduction

Recent years have seen a growth in water disputes linked to the Ganges, Brahmaputra and Meghna rivers. Collectively, these rivers form one of the largest water basins in the world, the Ganges–Brahmaputra–Meghna (GBM). The GBM is regarded as one of the least coordinated river systems in the world (ICA, 2012), in part due to its crossing of national boundaries between China, Bangladesh, Nepal, Bhutan and India (Khan, 1996; Bandyopadhyay and Ghosh, 2009). Some of these country's inhabitants are among the poorest people in the world.

With escalating population growth, rapid industrial development and increasing agricultural needs (Cronin *et al.*, 2014), India is experiencing a low level of water security, particularly over the management of the Indus and Ganges Rivers (Yan, forthcoming). One source of concerns lies in the pressing challenges that India is faced with in water management. India shares several rivers with its neighbouring states that include China, Bangladesh, Nepal, Bhutan and Pakistan. The shared river resources constitute India's main sources of water for domestic management (see Table 6.3). With an estimated gap of 50 per cent between water demand and supply (WRG, 2009), an increasing water demand is one of the country's largest challenges. As a developing state, India has shown ineffective management of its water resources, and is now facing increasing pressure to improve this management to help benefit the country's decelerating domestic economy (Biswas, 2011; Sahoo, 2014). Therefore, India has had disagreements with China, Bangladesh and other riparian states over the sharing of the GBM. India has been accused of exploiting the GBM's water resources to address the country's own national needs without negotiating with its neighbours. Concerns have been expressed about whether the riparian countries can survive such tensions without an escalation to military confrontations (Verma, 2013).

It has also been noted that developing countries often lack a commitment to the international environmental norms which facilitate interstate cooperation (Bernauer and Kalbhenn, 2010). To date, aside from bilateral agreements, no clear programme to develop collective water management practices has been initiated among the Brahmaputra's riparian countries. Although a water-sharing

*Table 6.1* Information relating to the GBM in its riparian countries

| *Countries the river runs through* | *River name* | *Length km* | *Watershed area km*$^2$ | *Rainfall mm/y* | *Evaporation mm/y* | *Runoff km*$^3$*/y* |
|---|---|---|---|---|---|---|
| China | Yarlung Zangbo | 1,905 | 345,083 | 1,145 | 459 | 121 |
| Bhutan | Brahmaputra | not applicable | 38,458 | 2,368 | 814 | 43 |
| India | Brahmaputra | 739 | 125,312 | 2,558 | 834 | 632 |
| Bangladesh | Jamuna | 272 | 23,215 | 1,949 | 673 | 679 |

Source: Sun *et al*. (2011).

agreement has been signed between India and Bangladesh, tensions have still arisen between these two countries. In the management of the GBM, neither India nor China is party to major global water norms,[1] and they both remain keen to cope with water conflicts through bilateral foreign relations.

Regional security in South Asia is considered to be unstable and rapidly changing, with the Sino-Indian relationship being a particular focus of discussions (Dahal *et al.*, 2003). As two neighbouring superpowers, India and China have had a troubled relationship. Territorial disputes are a key issue influencing China and India's bilateral relationship.[2] The two countries have been embroiled in constant clashes over disputed border areas. These have mostly been verbal clashes, but have sometimes escalated to small-scale military protests. Although progress has been made in resolving the border disputes, territorial boundary issues remain a potential source of conflict (Verma, 2013). An additional source of tension also lies in economic activities and trade, where the countries have developed close ties. China's trade with India has grown more rapidly than its trade with any of its top ten trading partners. In 2008, China surpassed the USA to become India's largest trading partner (Gupta and Wang, 2009). India has become China's biggest trading partner in South Asia, with a total trade exchange reaching US$65.47 billion annually in 2016. Nevertheless, a trade deficit has occurred for India and that has the potential to impact on the development of future Sino-Indian trade relations (Singh, 2013).

The water dispute is regarded as a source of conflict between India and China, which as large developing countries may experience challenges in resolving this conflict. Both are likely to be faced with concerns around water management, including the pressing needs to improve their water management capacity as well as development goals, and the requirement to improve economic, institutional and technological instruments (Biswas, 2011). The increasing water discords between China and India represent an intriguing issue in Sino-Indian relations which has strong implications for regional security.

This chapter aims to shed light on the nature, forms and outcomes of China's diplomacy over GBM. China's diplomatic initiatives have revealed vested interests and that the Beijing government is interested not only in seeking economic profits such as hydropower generation but the domination of hydrological data and water policy information as well. Both advantages and disadvantages exist in China's diplomatic initiative to provide hydrological data to India. With the change of government, India has vested its own interests in such interactions. With complicated dynamics that have developed in Sino-Indian water disputes, China's water diplomacy has not been a great success.

## China's policy over the Brahmaputra River

### *River features: security analysis*

The Brahmaputra River starts as a spring derived from the Angsi glacier in the Himalayas. As the Yarlung Zangbo River it runs eastwards across Tibet and then turns southwards at 'The Great Bend', turning again in South Tibet (where China borders India). From here, the Brahmaputra River travels west-south-west through north-eastern India. Along the way it is fed by many rivers on both its left and right banks, including substantial rivers that originate in Bhutan. Eventually, the river enters Bangladesh and flows south before joining the Ganges River to become the Padma River. The Padma River flows south-east to join the Meghna River which flows downstream in a southerly direction to finally reach the Bay of Bengal.

The extent of water quantity competition is different between India and China and between India and Bangladesh, due to the character of the GBM river basin, which has a large spatial disparity in precipitation. Since the majority of the region's annual rainfall occurs during the monsoon season, which lasts about two-and-a-half months, a wide temporal inequity exists (Bandyopadhyay and Ghosh, 2009). In addition, the drainage area of the GBM differs depending on how India and China's disputed border is drawn. Two sets of analysis exist illustrating that water usage within Chinese territory does not affect water availability in the lower stream in India. The border outlined by the Chinese government shows that in relation to the GBM's total drainage system, only a small percentage of the drainage area lies within China. The data becomes negligible when the Indian government's official borderline is adopted (Rahaman, 2009). Therefore, this study will use China's demarcation of the border to show the largest impact China may potentially have on the lower stream of the GBM. From this viewpoint, water from China's mainstream Yarlung Zangbu River contributes 14.61 per cent of the total annual runoff of the Brahmaputra River. River runoff of the Ganges delta drainage system is mainly generated in India, Nepal, Bhutan and Bangladesh (91.12 per cent), and the contribution of water from China is small in terms of the total natural runoff (8.88 per cent). Therefore, water from China has little impact on the development of water resources in downstream countries, including India, Bangladesh and Bhutan (see Tables 6.1, 6.2 and 6.3, and Figure 6.1) (Sun *et al.*, 2011). Accordingly, China's water usage at the upper reaches of the Brahmaputra River is less than 3 per cent of local surface water resources and less than 0.25 per cent

*Table* 6.2 The percentages of the three rivers' natural runoffs from China

| *River name* | *Annual runoff ($km^3/y$)* | *Annual runoff in China ($km^3/y$)* | *Percentage of China to the total (%)* |
|---|---|---|---|
| Ganges | 550 | 14 | 2.55 |
| Yarlung Zangbo/ Brahmaputra | 828 | 121 | 14.61 |
| Ganges delta (Ganges–Brahmaputra–Meghna River) | 1,520 | 135 | 8.88 |

Source: Sun *et al.* (2011).

*Table* 6.3 China, Bhutan, India and Bangladesh water resource management in the Yarlung Zangbo River

| | *River catchment area ($km^2$)* | *Runoff depth (mm)* | *Water resources ($10^9 m^3$)* | *Agricultural water withdrawal ($10^9 m^3$)* | *Industrial water withdrawal ($10^9 m^3$)* | *Domestic water withdrawal ($10^9 m^3$)* | *Total withdrawal ($10^9 m^3$)* | *Basin population* | *GDP per capita on the river (US$)* |
|---|---|---|---|---|---|---|---|---|---|
| Bangladesh | 76,834.22 | 1,604.27 | 1,232.6300 | 215.3851 | 27.4431 | 5.1142 | 247.9423 | 109,865,091 | 75.6 |
| Bhutan | 39,835.11 | 1,328.89 | 529.3645 | 3.0326 | 1.4744 | 0.0343 | 4.5413 | 2,739,207 | 54.8 |
| China | 34,437.60 | 968.05 | 3,334.3238 | 22.0414 | 1.0559 | 1.5633 | 24.6606 | 2,754,981 | 667.6 |
| India | 172,292.31 | 2,512.14 | 4,328.2283 | 80.3318 | 15.3677 | 6.9421 | 102.6415 | 47,089,712 | 87.6 |

Source: Yan (forthcoming).

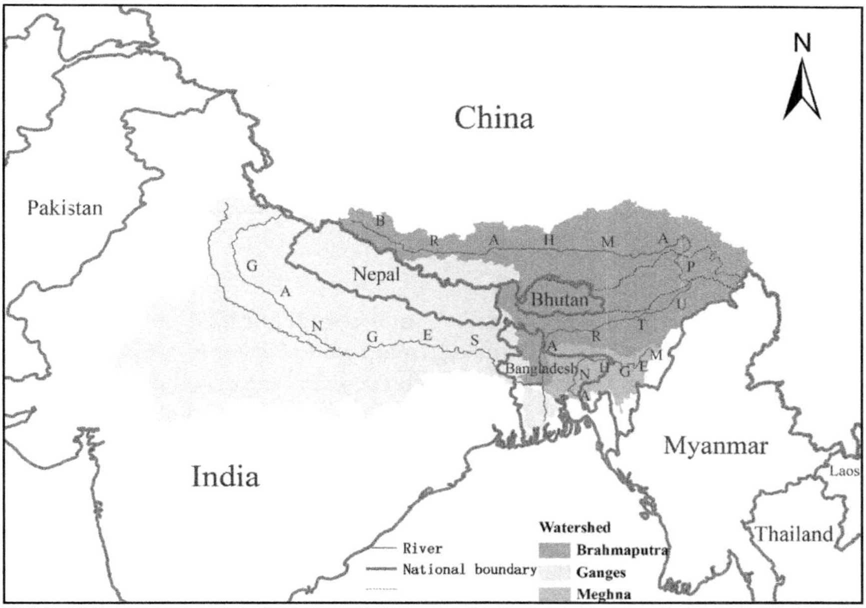


*Figure 6.1* The geopolitical units in the Brahmaputra–Meghna River basin.
Source: authors' commission work, credit: Wenbin Zhu.

of the total water resources of the Ganges delta drainage system. It is therefore evident that water usage within China has very little impact on the lower reaches of the GBM.

This analysis suggests that in terms of water quantity competition, India's relationship with Bangladesh is different from its relationship with China. First, India competes fiercely with Bangladesh for limited water supplies during the dry season, thus seriously challenging India's water management capacity. Meanwhile, in its relationship with China, India is more focused on the ecological impact on Arunachal Pradesh from dam construction projects in China's territory. There is a heightened risk of flash flooding during the monsoon season, which poses a threat to Indian residents along the GBM. In contrast to its relationship with China, India's water usage has a greater impact on Bangladesh during both the monsoon and dry seasons, affecting not only Bangladeshi society and the environment, but also the nation's economy. In particular, one study indicates that the diversion of the Ganges at Farakka has negative socio-economic and environmental impacts during the dry season. The Farakka Barrage has affected agriculture, fisheries, industry, navigation, forestry and native vegetation in the south-west region, which comprises about a quarter of Bangladesh's landmass (Rahaman, 2009). Bangladesh suffers water-related problems to a much larger degree of magnitude than India suffers from any similar problems caused by China.

*Table 6.4* Ratio of water resource use to the GBM

| | *Yarlung Zangbo River (%)* | *Ganges River (%)* | *Indus River (%)* | *Average security level on the GBM (%)* |
|---|---|---|---|---|
| China | 0.7 | 0.6 | 0.1 | 0.5 |
| India | 2.4 | 110.5 | 109.2 | 74.0 |
| Bangladesh | 20.1 | 37.2 | not applicable | 28.7 |

Source: Yan (forthcoming).

With regard to the three countries' water resource utilization rate, China, India and Bangladesh have displayed different levels of water insecurity over the GBM (see Table 6.4). China is most secured. Bangladesh is the second most secured, although its use of the Ganges' water resources has reached a high level in proportion to that of the total river basin. India, in comparison, is the least secured in water resource utilization. It has reached the full potential for water resource use from the Ganges and Indus Rivers, although large potential still exists for use of the Yarlung Zangbo River, which is located in the north-east of the country. Therefore, India suffers a high level of water insecurity in the GBM.

### *China's policy focus on the GBM*

China's commitment to cooperative agreements is boosted in its intention to construct dam projects including the Zangmu Dam (藏木) and the Laohuzui Dam (老虎嘴). The country is focused on harvesting economic benefits through resource governance. The Brahmaputra River system is viewed as 'under exploitation' by both China and India (Jia *et al.*, 2014), which is not surprising given that further development plans have been made to extract economic benefits from it. In both the 11th and 12th Five-Year Plans (2006–2010 and 2011–2015), provision was made for the exploitation of hydropower in south-west China including the Zangmu Dam and those proposed on the Yalong, Dadu and Nu Rivers. These projects were all contentiously protested against. The Zangmu Dam, which generates 510 MW and is located on the middle reaches of the Yarlung Zangbo River, started to operate in 2014. The total capacity of China's hydropower has now exceeded 225 GW, and the contribution of the Zangmu Dam in particular is equivalent to the entire previously existing hydropower-generating capacity of Tibet. Nevertheless, the total generation capacity of the region is estimated at approximately 3,000 MW.[3] Dam constructions on transboundary tributary rivers including the Yarlung Zangbo and Irrawaddy Rivers are also being planned or developed. Among these, the Lalho project on the Xiabuqu River, a tributary of the Yarlung Zangbo and Brahmaputra Rivers, commenced in June 2014. The Jitaiqu or Kelaolong dam project, a tributary of the Irrawaddy River, started in late 2017.[4]

Following the Chinese government's interest in meeting its economic goals, it has extended its domestic water resource management policies to cover all rivers that run through its national borders. A coherent set of policies to plan and manage water resources has been developed in order to meet the targets set out in the Eleventh and Twelfth Five-Year Plans,[5] which deem the building of the Zangmu Hydropower Station on the Brahmaputra River as an internal affair. Hence, details of its planning, construction and operation were not disclosed to the Brahmaputra River's downstream riparian states.

## India's policy and principles over the Brahmaputra River

India shares transboundary rivers with its neighbouring countries, including China, Bangladesh, Pakistan, Nepal and Bhutan. From domestic management and foreign policy perspectives, India's management of transboundary rivers has shown a number of features that largely contradict its promotion of the sustainable management of its shared water resources.

### *Domestic water management*

India requires improved domestic water management to advance the country's economic, institutional and technological development (Biswas, 2011). With regard to water ownership, state governments are entitled to own water resources and make decisions over water management. However, the Indian central government has the discretion to intervene in local policies. Lacking an absolute authority, water policies from different states may contradict each other. These policies, and associated on-ground water management, are therefore not optimized at the scale of water basins. Such a disjointed system has resulted in the degradation of ecosystems along India's international rivers (Das, 2011; Kattelus *et al.*, 2015; Jia *et al.*, 2016).

Matters on transboundary river management are a highly sensitive political issue in India, and are held as state secrets. Information that would negatively affect national sovereignty or international relations, including policies for managing the rivers and hydrological data for the planning of water resources (such as stream and sediment flow, water withdrawal, and usage) is notoriously difficult to access within the country (Prasai and Surie, 2015). While data from non-classified sites such as India's peninsular rivers is published in a yearly hydrological data book, similar micro-level data for transboundary rivers (including the Ganges, Brahmaputra and Indus Rivers) is not disclosed (Seth, 2013). At the same time, the country's planning and management for its transboundary rivers often suffers from a lack of hydrological data, or there is only a piecemeal and uncoordinated gathering of relevant information by government authorities (Prasai and Surie, 2015).

India's water management policymaking processes lack public participation and as a result, state–society tensions have the potential to lead to interstate tensions and conflicts. Increasing environmental activism has questioned the

Indian national government's policy to maximize economic profits from water resources such as through dam constructions (Grumbine and Xu, 2011) and water diversion projects (*inter alia* Chowdhury, 2010; Sood and Mathukumalli, 2011). Nevertheless, the current institutions have not yet guaranteed the incorporation of the public in the policymaking arena. In many cases, Environmental Impact Assessment laws have been the primary legal document adopted to defend citizens' environmental rights (Grumbine and Pandit, 2013). Clashes have occurred when central authorities promote the exploitation of water resources without compliance with environmental laws or relevant legal regulations (Nandimath, 2007). In many cases, the absence of legitimate channels for the articulation of grassroots opposition has unfortunately led to violence and radical actions (Huber and Joshi, 2015).

### *India's diplomatic practices in water cooperation*

India has developed varied approaches to water diplomacy with its riparian neighbours. However, this has not been a smooth process and India has been accused of exploiting water resources to address its own national needs without negotiating with its neighbours. In most cases, India has adopted a 'bilateral cooperative' approach to securing its interests in water resource management. India has developed bilateral cooperation with most of the countries that it shares water resources with (see Table 6.5). In the absence of sanctions, the country's attitude towards its upstream and downstream neighbours differs. With countries located on the upstream of the shared rivers, including Pakistan and Nepal, India has shown willingness to compromise by sharing policy information. In comparison, with downstream countries, India has taken unilateral initiatives to manage water resources within its own territory (Rahaman, 2009).

India and Bangladesh have managed to form institutionalized arrangements in recent decades, but disagreement remains. For Bangladesh, water scarcity is an essential concern. One focal point of the disagreement between the two countries is the sharing of water from the Ganges River during the dry season (December to May). India has established treaties regarding the mechanisms

*Table 6.5* India's bilateral mechanisms with its riparian neighbours

| *Country of interaction* | *Treaty* | *Water information sharing* |
|---|---|---|
| Bangladesh | The 1996 Interim Agreement on the Ganges | None |
| Nepal | The Kosi Agreement (1954)<br>The Gandak Agreement (1959)<br>The Tanakpur Agreement (1991)<br>The Mahakali Treaty (1996) | Policy information |
| Bhutan | None | None |
| Pakistan | The Indus Waters Treaty (1960) | Policy information and hydrological information |

and principles of water division with Bangladesh, including a 1977 agreement and two memoranda of understanding (MOUs) in which information disclosure was mentioned.[6] Although information disclosure was mentioned in the 1983 MOU, it was dropped in the subsequent 1996 agreement. This agreement established that a Joint Rivers Committee was to collect water information to record flows at Farakka (Rahaman, 2009). There is no mention of policy information sharing from India when initiating projects to manage water within the country. In the absence of sanctions, there has been little compliance with the water agreement (Price, 2014). Moreover, senior Bangladeshi officials are dissatisfied with India's sharing of information (Price, 2014), and claim that Bangladesh has never received the water it is due under the 1996 treaty (Ramachandran, 2006). India's project to build domestic river links has raised further disapproval both within India and Bangladesh. The Bangladeshi authorities have, on various occasions, appealed to the global community and sought international interventions, including at the UN and the Commonwealth Summit (Khalequzzaman, 2006; Rahaman, 2006). Over recent years, military tensions have arisen over the supply of irrigation water from Teesta (Price, 2014). The frequency of warnings coming from Bangladesh indicates that it is on the verge of escalating its water conflict with India (see Table 6.6).

### *India's responses to China's water diplomacy*

The evolving water policy in India has provided important insights for understanding the country's position towards China's water diplomacy. One cause of concern for India is China's unilateral planning and use of water resources from the Yarlung Zangbo River. As no formal agreement over the Zangmu Dam/Yarlung Zangbo River has been reached between the two countries, Indian elites and mass media have begun to raise concerns about China's reluctance to provide relevant information. In particular, it is feared that China's upstream water diversions and usage will have a detrimental impact on India. According to such a view, China as the upstream stakeholder was single-minded in its use of water and may consider it as a weapon to obstruct India's development. With the river diverted and prevented from flowing into India, the livelihoods of the people who live and depend on the lower stream of the river will be affected.[7]

Different Indian governments have shown rather different attitudes towards China's management of shared water resources. Disaster control was an essential concern in the early years under the Singh government (2004–2014). During the monsoon season, there are significant risks of flooding which threatens residents along the Brahmaputra River basin in India's north-eastern region of Arunachal Pradesh. China's unilateral planning and use of water resources from the Yarlung Zangbo River has further enhanced such concerns for India. Since 2000, flash floods have caused large-scale destruction along downstream parts of the GBM in Arunachal Pradesh. One of the key concerns was China's launching of water diversion projects on the Brahmaputra/Yarlung Zangbo River. It is argued that China's water diversion projects in the western bend of the Yarlung

*Table* 6.6 India–Bangladesh political interactions over water

<table>
<tr><th></th><th>India–Bangladesh</th><th>Bangladesh–India</th></tr>
<tr><td>1951</td><td></td><td>Pakistan objects to India constructing the Farakka Barrage</td></tr>
<tr><td>1961</td><td>India admits to the unilateral construction of the Farakka Barrage</td><td></td></tr>
<tr><td>1964</td><td></td><td>Master plan for water resources development-flood protection</td></tr>
<tr><td>1972</td><td colspan="2">Signing of Indo-Bangladesh Joint Rivers Commission</td></tr>
<tr><td>1974</td><td>Joint statement for studying and research</td><td></td></tr>
<tr><td>1975</td><td colspan="2">Bangladesh's suggestion to increase diversion of water not accepted</td></tr>
<tr><td>1976</td><td></td><td>Opposes the Farakka Barrage through regional multilateral cooperation</td></tr>
<tr><td>1977</td><td colspan="2">Both sides sign Ganges Water Agreement</td></tr>
<tr><td>1982</td><td colspan="2">Both sides sign MOU for sharing dry season water flow from the Ganges River</td></tr>
<tr><td>1985</td><td colspan="2">No execution of 1982 MOU; no water sharing; MOU expires in 1988</td></tr>
<tr><td>1993</td><td></td><td>Raises the issue in Cyprus at the Commonwealth Summit</td></tr>
<tr><td>1995</td><td></td><td>Raises the issue</td></tr>
<tr><td>1996</td><td colspan="2">Both sides update the Ganges Water Treaty for the duration of 30 years</td></tr>
<tr><td>2005</td><td></td><td>Raises the issue of augmenting the dry season flow by building reservoirs in Nepal</td></tr>
</table>

Zangbo River may lead to water shortages in India's north-eastern regions, whereas excessive water releases in China, as a 'dam controller' mechanism, may lead to floods. For this reason, Brahma Chellaney, an influential scholar in India, claims that water has become China's weapon against India.[8]

Accordingly, in its diplomacy with China, India has shown strong concern about sharing hydrological information so that it can monitor the ecological impacts of upstream damming on the GBM. As China has conformed to India's request, the Indian central government further attempted to respond to the public's distrust over China's dam construction. The then Prime Minister Singh and the External Affairs Ministry reacted in a calm fashion consistent with the official position on the matter, persuading the Indian public that correct judgements were made about the significance of water competition with China: 'We trust but we also verify. We have verified Chinese claims on the dam being a run-of-the-river power project which does not store water. We are convinced that it is a run-of-the-river project.'[9] Official concerns over water sharing have also evolved in more recent years. Under the Modi government (2014–), unlike the Singh government, a populist government started to rise. India's policy over transboundary rivers became incorporated into its foreign policy agenda. Concerns over the Brahmaputra River, such as flooding, were originally a domestic issue that has affected north-east India's economic conditions. After Modi assumed government, the importance of this issue was heightened due to increasing political pressure to eliminate challenges in water management. To boost the newly elected government's popularity, Modi reinforced the elite's questioning of China's dam construction (Panda, 2013). Unlike its predecessor, the Modi government has raised concerns about transboundary rivers with China. In his opening address during a trip to China in 2015, Modi stated: 'I stressed the need for China to reconsider its approach on some of the issues that hold us back from realizing full potential.... I sought tangible progress on issues relating to visa policy and transboundary rivers'.[10] These statements highlight that the matter of transboundary rivers has become closely linked with bilateral issues that represent ongoing concerns for India, and that the Modi government has hardened its position.

The Modi government's attitude towards China is also reflected in its pushing through of national water management projects. Damming the Brahmaputra River had been on the agenda for the Singh government. However, as a result of local protest and transnational environmental activism, the projects were halted. In order for these policies to be legitimized, the Indian central government has also tactically reformed water management institutions to facilitate the decision-making process on the damming project (see Box 6.1). Additionally, the Modi government has advocated that China's damming activities have negative impacts over the north-eastern part of India.[11] It has politicized the importance of the proposed damming projects as an illustration of the exercising of its rights as a riparian state on the transboundary river. By so doing, Modi had distracted the public's attention by emphasizing that China is an obstacle to India's rightful appropriation of its water resources.

**Box 6.1 Dam construction in Arunachal Pradesh under the Modi government**

The Modi government has proposed to build a dam over a tributary of the Brahmaputra River in Arunachal Pradesh. The planned project is estimated to have a capacity of 9–13 billion $m^3$ in the Siang region of the state. The dam is said to help control floods in Assam and to generate hydroelectricity.

Such plans have, however, been protested by environmentalists and local residents who opposed large dam constructions. The Brahmaputra Board is another force that has discretion over the monitoring and approval of dam projects.[12] It is an independent government unit under the Ministry of Water Resources, and is constituted by all jurisdictions of the north-eastern region. It aims to coordinate all affected states' interests. On dam construction plans, it has placed focus on flood alleviation and insists that the large scale of the proposed dams has the potential to cause ecosystem damage. The Board has, however, gradually lost its authority as the central government has tightened its control over water planning and management. In order to promote its dam construction, the government has attempted to restructure the Brahmaputra Board. The central government, led by minister Uma Bharti, has made attempts to restructure the Board by weakening its authority as an expert in flood alleviation while focusing on harvesting profit from water resources.[13] Nevertheless, such attempts did not succeed as a result of the Assam State government's opposition.

## China's water diplomacy

China's diplomatic initiative with India had progressed stably in the early 2000s since it had started to plan and manage water resources of the Yarlung Zangbo River. In response to flooding in India in 2000, and having closely monitored the situation for weeks beforehand, the Chinese authorities sent warnings to the Indian government about the likelihood of landslides. This move helped the two countries develop water diplomacy around the issues of 'natural disasters'[14] and 'emergency management'.[15] Subsequent talks on the issue between the two nations led to an agreement over the Brahmaputra River. In 2002, the two countries agreed on an MOU in the form of a bilateral agreement, which has been renewed in the years since it was formed (see Table 6.7). These progressive diplomatic initiatives have been effectively implemented. The MOUs, in the form of hydrological information sharing, indicate that despite a low level of institutionalization, agreement had been reached to promote water dispute resolution.

Continual joint ministerial meetings have been held, thus indicating that diplomatic cooperation has been established. After a few years of China unilaterally providing information, a joint Expert Level Mechanism (ELM) was also established at the ministerial level in 2006.[16] Between 2006 and 2014, eight meetings of the ELM were held for the purpose of helping India to use the data it was receiving to develop effective flood forecasts.[17] The mechanism was

*Table 6.7* Political interactions between India and China over water and other issues

| *Date* | *Bilateral agreements on water management* |
|---|---|
| 2000 | Flash floods caused large-scale destruction in the lower stream of the Brahmaputra River in Arunachal Pradesh. Subsequent talks on the issue between the two nations led to an agreement over the Brahmaputra River. |
| 11 April 2005 | MOU signed with China confirms provision of data on one more location, hydrological data from China in respect of two rivers – the Brahmaputra and the Sutlej/Langen Zangbu. |
| 2011 | For provision of hydrological information of Sutlej and Langen Zangbu rivers in the flood season by China to India with a validity of five years. |
| 2013 | China agreed to extend its provision of hydrological data.[1] |
| 2014 | India confirms China's compliance with water information sharing agreement. |

Note

1 It was agreed that water data was to be provided from 15 May to 15 October every year on a daily basis, adding 15 days to an earlier agreement.

primarily composed of state-sponsored experts and hydrologists, to guarantee that the bilateral cooperation excluded the involvement of experts who have no affiliations to either state.

Hydrological data is the main form of water information shared between the two countries. Gradual progress was made between 2002 and 2014 in the volume of hydrological data provided. In 2005, China agreed to unilaterally provide data on the Sutlej and Langen Zangbu Rivers during the flood season for up to five years. Further, in 2013, China agreed to extend its provision of daily hydrological data from 15 May to 15 October every year, adding 15 days to an earlier agreement. As requested by the Indian government, water information during the monsoon season is now provided. With continuous dam construction projects planned on the upper stream such as Zangmu Dam and Lalho project, China has gradually expanded the scope and time period of the data provided. The MOUs are recognized by the Indian government, which in 2013 formally acknowledged that China's dam construction had no direct impact on water quantity in India. Accordingly, China has unilaterally provided water information during the monsoon season in a bid to eliminate India's concerns about water quantity and flooding in the Brahmaputra River basin.

Reflecting the evolution in India's approach to water diplomacy, the Modi government has made a tentative request for China to provide hydrological data all-year round. This would help India to monitor China's upstream plans and management of the GBM. China declined the request and has shown no intention to disclose its plans for the country's future management of the river. To China, the agreed sets of data have made it technically possible for India to predict water disasters and, hence, should serve to build trust between the two countries. Modi's move is probably deemed by China as a form of provocation and an obstacle to the two countries' pursuit of cooperation over international water management. And China had not proceeded with the plan to share more data that are collected on the Yarlung Zangbo and Lohit Rivers.[18] By so doing, China has stopped short in expanding cooperative initiative with India by providing hydrological data over the recently built dams.

### *Complex nature of China's water diplomacy*

Due to the Zangmu Dam's geographical location in the upstream of southern Tibet, it is not surprising that China's exploitation of water resources is linked to interests surrounding border disputes (see also Huang, 2015). Along the Sino-Indian border there are approximately 125,000 km$^2$ of disputed territory, partly falling within southern Tibet. As the Brahmaputra River runs through the eastern part of the southern Tibet, concerns over water-sharing for this river have thus become more complex, as they are intertwined with the border disputes and the seeking of independence by Tibet.

#### *The issue of Tibet*

The Brahmaputra River is sourced from the Himalayan Mountains and runs through the disputed area of the Arunachal Pradesh region. India claims that Tibet's water is a 'common resource' and this is regarded as provoking China. As expressed in the *Global Times*, an official national Chinese newspaper, the concerns over China's dam construction projects raised by the Indian media have 'put more pressure on China by exaggerating the facts and drawing attention from the international community, with the intention of preventing China from developing Tibetan water resources' (Li, 2013). The different border alignments adopted by India and China therefore further complicate the understanding of the Brahmaputra River as a holistic system,[19] and have stagnated the two countries' diplomatic cooperation over the sharing of the GBM.

A prominent Chinese perspective on the issue of sharing the GBM is that India is trying to manipulate how China manages water resources in the region by building influence in Tibet.[20] This suspicion was enhanced when the Dalai Lama emphasized the significance of these water resources as those of Tibet when he warned of looming dangers stemming from the Tibetan plateau.

> The ecology in Tibet ... is something very, very essential. So, since millions of Indians use water coming from the Himalayan glaciers ... I think you (India) should express more serious concern. This is nothing to do with politics, just everybody's interests, including Chinese people.[21]

Accordingly, without China's agreement not to act unilaterally on the GBM, another concern lies in a water diversion project on the Tibetan plateau, from southern to northern China, also known as the South–North Water Diversion Project. The water is to be taken from the Yarlung Zangbo River. This project was heatedly discussed in 2011 and raised fears in India that water flow would be substantially affected in the Brahmaputra (Panda, 2016).

### *Inter-issue negotiation*

It is worth noting that since China initiated its water diplomacy, China and India's water cooperation has been closely combined with talks on border dispute resolution (Panda, 2015). And China has displayed an intention to deliberately link the water-sharing issue with territorial matters. During key events in China's water diplomacy, including in 2005, 2006 and 2013, water data-sharing agreements have served as key bargaining tools for China in its efforts to reach agreements over border disputes (see Table 6.6). Since 2005, Sino-Indian cooperative activities have become institutionalized in the form of China's unilateral dissemination of hydrological information. In the same year, to address disagreement on the border dispute, both countries signed The Strategic and Cooperative Partnership and Agreement on Political Parameters and Guiding Principles for the Settlement of the India–China Boundary Question (2005). Although this agreement over the border dispute may not prevent conflicts, China believes that it will help to de-escalate tensions and avert military face-offs along the Line of Actual Control (Stobdan, 2013). However, the Indian side believes that these benefits serve only China's interests and potentially support its military intrusion into disputed areas (Stobdan, 2013).

## Assessment of China's water diplomacy

China's water diplomacy over the GBM is primarily dominated by delicate Sino-Indian foreign relations. Although the country has developed institutional cooperation with Bangladesh, such interactions are far less than those between both China and India, and Bangladesh and India. As Sino-Indian water diplomacy is mostly centred around institutional agreements on data exchange, the evolution of such institutional cooperation is of importance to understand the outcomes of China's water diplomacy. This section assesses the significance of such cooperation by first reviewing the relevant literature.

According to scholarship focusing on agreements over shared water resources, information disclosure represents a key international norm to mitigate conflicts (Stinnett and Tir, 2009; Mitchell and Zawahri, 2015). It is noted that various types of agreements for information-sharing exist. Some are deliberately designed to vaguely relate to data exchange. According to Fischhendler (2008), such purposeful vagueness indicates negotiators' intentions to serve domestic political purposes. He argues that in order to serve domestic interests, countries are inclined to refuse to provide requested data, or to deliberately negotiate vague terms in the agreements. Furthermore, the exchange of water data can provide opportunities for transparency and mutual assurance (Burton and

Molden, 2005; Gerlak *et al.*, 2011). It promotes communication and understanding between neighbouring riparian states (von Stein, 2008). The activity is likely to mitigate distrust and enhance confidence in countries forming a collaboration. Networks are likely to be formed by implementing data exchange (von Stein, 2008), and the contact communication also has the potential to lead to the development of joint scientific research (Uitto and Duda, 2002). Hence, hydrological data can serve as a primary condition for formal negotiations to take place (Conca *et al.*, 2006; Karkkainen, 2006; Zawahri, 2008, 2009). These negotiations can lay the foundation for broader agreement on transboundary water management (Miles *et al.*, 2002; Uitto and Duda, 2002), and for sustainable policy solutions to be sought (Dimitrov, 2006).

Disadvantages exist in countries' forming water data-sharing agreements. Although data sharing may facilitate beneficial communications and coordinated water management, it also has the potential to obstruct negotiations and impede further cooperation. In contrast to flexible arrangements of water data-exchange schemes, strict and more precise water data agreements may propel states to fulfil their obligations (Abbott and Snidal, 1998; Gerlak *et al.*, 2011). This type of agreement, compared to flexible agreements, is less likely to succeed in negotiations. Some riparian states may persist in seeking data or information without reasonable cause (Timmerman, 2004; Gerlak *et al.*, 2011). In particular, under the current international legal framework, downstream countries are 'facilitated' to 'harm' upstream countries by requesting water information (Salman, 2010). In such situations, water data has become a weapon to harm other parties.

This literature review helps us to understand the significance of China's information-sharing agreement with India. For some time, water diplomacy between the two countries has progressed relatively positively. Sino-Indian water cooperation has benefited from information exchange. Both countries have developed common concerns regarding water management, which have resulted in the formation of hydrological data exchanges. Indeed, the negotiations for the information-sharing MOUs progressed smoothly, with China showing a willingness to expand the scope of the hydrological data it could provide. In particular, monitoring data during the monsoon season contributes to disaster control, although only partial information is shared. Implementation of the data-sharing agreement has also been smooth. This mechanism also loosely serves as a symbolic institution for managing conflicts. It is believed the two countries' joint ELM can lead to better understanding of the data for the Indian authorities to formulate flood forecasts.[22]

However, in Sino-Indian water cooperation, benefits from such agreements were minimal and were soon replaced by distrust that challenges the scope of information that China provides. For China, it is apparent that the country's signing of water information-sharing agreements indicates that the country is only willing to provide partial water data in exchange for broader foreign relations benefits. This is because the agreement is purposefully designed to serve China's self-interests.

It is evident that India has also vested its own interests in forming institutional cooperation over water management, indicating a different incentive structure from that of China. The MOUs between China and India serve to monitor China's management of the GBM and to ensure that India's interests are not harmed. Along with China's unilateral provision of water information, and other signs indicating trust in the two countries' foreign relations, India not only dismissed China's initiative but has also pushed further with provocation.

Therefore, the agreements' performance remains limited, despite that it may be helping India to formulate flood forecasts. The purpose of such water information-sharing agreements for China is to maintain the non-disclosure of its policy information on the planning and management of the GBM. Consequently, diplomatic cooperation has barely served to promote collaboration over water resource management.

## Concluding remarks

China and India have achieved a limited degree of formalization in their initiatives for managing their shared rivers, with only preliminary cooperation being formed. Although institutional cooperation has been formed between China and India, it is also recognized that the water-sharing agreements satisfy both countries' interests and are intertwined with other Sino-Indian relations and foreign policy issues. For a short while, China was relatively satisfied with the benefits of the agreement, including improved international trust-building and the country's successful dam construction projects (such as the Zangmu Dam) that were undertaken without formal negotiations. Overall, the rationale for China's water diplomacy is not foremost to resolve issues around water management, but instead to use water as a bargaining tool in bilateral relations.

Water has become a politicized foreign policy issue in India, as reflected in the government's evolving position on the matter. India's approach under the Modi government, in particular, indicates that upholding the country's strategic interests and attracting voters in future elections have been prioritized. This approach sought to legitimize the country's position by adopting the rhetoric of a downstream riparian state demanding justice and equality over transboundary water management.

Given these mismatched interests and approaches, Sino-Indian diplomatic initiatives over shared water resources have entered a deadlock. This is currently blocking collaboration, and further preventing both parties from fully benefiting from the GBM river basin. Without substantially improved diplomatic interactions over key foreign policy issues, the water institutions that can lead to effective management of the two countries' shared rivers will not be established. More specifically, this situation could hinder India's ability to develop hydropower facilities on the GBM rivers. It has also greatly affected the interests of other riparian countries in the region, including Bhutan, Nepal and Bangladesh. Indeed, some policy adjustment can be seen in China's water diplomacy with Bangladesh, another downstream country. The risks that both China and India are taking could be heightened, considering their historically unstable relations. No formal agreement has succeeded in determining how the two countries will jointly plan and manage the water resources on the GBM. A formal dialogue process or agreement about issues relating to the GBM is greatly needed. There is little sign of this at present, given the recent deterioration of Sino-Indian water diplomacy. It appears unlikely that either country will take the initiative to develop collective water management over the GBM river basin.

## Notes

1 These include the UN Convention on the Law of Non-Navigational Use of International Watercourses (1997); the UNECE Convention on the Protections and Use of Transboundary Watercourses and International Lakes (1992); and the Helsinki Rules (1966).

2 The two countries have two major points of border dispute. One, in the western sector, is over Aksai Chin in the north-eastern part of Jammu and Kashmir in India. India has alleged that China is occupying 38,000 $km^2$ of its territory in Kashmir. The second dispute, in the eastern sector, is over part of the Arunachal Pradesh area which China claims to be part of South Tibet. China asserts that Arunachal Pradesh is an integral part of the country.

3 Meng, X. It is time to exploit hydropower electricity in Tibet, *People.cn*, 29 November 2010. Available at: http://paper.people.com.cn/zgnyb/html/2010-11/29/content_684332.htm (accessed 25 July 2017).

4 Tibet increases the exploitation of hydropower development, *Renmin Net*, 17 July. Available at: http://paper.people.com.cn/zgnyb/html/2017-07/17/content_1791768.htm (accessed 17 July 2017).

5 Personal interview with a scientist from the Chinese Academy of Sciences.

6 These are the 1983 and 1985 MOUs.

7 Brahma Chellaney, 'China's hydro-hegemony', *New York Times*, 7 February 2013. Available at: www.nytimes.com/2013/02/08/opinion/global/chinas-hydro-hegemony.html (accessed 5 August 2015).

8 Brahma Chellaney, 'Leader article: Counter China's Designs', *The Times of India*, 16 January 2008. Available at: http://articles.timesofindia.indiatimes.com/2008-01-16/edit-page/27755750_1_tibet-and-burma-gwadar-india-and-china (accessed 25 July 2017).

9 Mehdudia, S. We trust China's promise on dam: Manmohan, *The Hindu*, 5 August 2011. Available at: www.thehindu.com/news/national/we-trust-chinas-promise-on-dam-manmohan/article2322266.ece (accessed 5 September 2016).

10 Prime Minister's Media Statement in Beijing during his visit to China (15 May 2015), Ministry of External Affairs, Government of India, 15 May 2015. Available at: www.mea.gov.in/Speeches-Statements.htm?dtl/25239/Prime_Ministers_Media_Statement_in_Beijing_during_his_visit_to_China_May_15_2015 (accessed 29 November 2015).

11 See, for example: www.news18.com/news/india/china-and-pakistan-not-letting-india-use-its-own-resources-1301293.html; and www.northeasttoday.in/modi-government-not-serious-about-the-issues-of-assam-says-gogoi/ (accessed 25 July 2017).

12 According to the Brahmaputra Board Act, the mandate of the board is to prepare plans and implement integrated measures for flood control and the prevention of river bank erosion in the Brahmaputra valley.

13 Kalita, P. 'Failed' Brahmaputra board to be restructured, *The Times of India*, 20 June 2015. Available at: http://timesofindia.indiatimes.com/city/guwahati/Failed-Brahmaputra-board-to-be-restructured/articleshow/47746796.cms

14 On 23 October 2013, the Border Defence Cooperation Agreement (BDCA) was signed between China and India.

15 'Joint Statement between the Republic of India and the People's Republic of China on Building a Closer Developmental Partnership', Ministry of External Affairs, 19 September 2014. Available at: www.mea.gov.in/bilateral-documents.htm?dtl/24022/Joint+Statement+between+the+Republic+of+India+and+the+Peoples+Republic+of+China+on+Building+a+Closer+Developmental+Partnership (accessed 19 November 2015).

16 The Indian side is led by the Commissioner, Ministry of Water Resources, while in China it is led by the Director, International Economic and Technical Cooperation and Exchange Center, Ministry of Water Resources.

17 *India–China Co-operation*, Ministry of Water Resources, River Development and Ganga Rejuvenation, Government of India. Available at: http://wrmin.nic.in/forms/list.aspx?Id=4&lid=349 (accessed 24 August 2017).

18 During the 2005 ELM meeting, China has indicated that it was possible to provide hydrological data on the two sites of Yarlung Zangbo and Lohit Rivers.

19 According to the national border demarcated by the Indian government, the river basin has a minor branch that runs through South Tibet in China.

20 India pledges for a Lhasa Consulate turned down by China, *Global Times*, 19 April 2014. Available at: http://news.ifeng.com/a/20140419/35892642_0.shtml (accessed 25 July 2017).

21 Tibetan glaciers melting, Dalai Lama claims, *Huffington Post*, 4 February 2011. Available at: https://web.archive.org/web/20110403054927/www.huffingtonpost.com/2011/04/02/tibetan-glaciers-melting-dalai-lama_n_844002.html (accessed 25 July 2017).

22 India–China Co-operation (A) Memorandum of Understanding Upon Provision of Hydrological Information of the River Brahmaputra/Yarlung Zangbo, Ministry of Water Resources, River Development and Ganga Rejuvenation, Government of India, 7 June 2016. Available at: http://wrmin.nic.in/forms/list.aspx?lid=349 (accessed 25 July 2017).

## References


Abbott, K.W. and Snidal, D. 1998. Why states act through formal international organizations, *Journal of Conflict Resolution*, 42(1): 3–32.

Bandyopadhyay, J. and Ghosh, N. 2009. Holistic engineering and hydro-diplomacy in the Ganges–Brahmaputra–Meghna Basin, *Economic and Political Weekly*, 44(45): 50–60.

Bernauer, T. and Kalbhenn, A. 2010. The politics of international freshwater resources, in R.A. Denemark, ed., *The International Studies Encyclopedia*, Malden, MA: Wiley-Blackwell, pp. 5800–5821.

Biswas, A. 2011. Cooperation or conflict in transboundary water management: Case study of South Asia, *Hydrological Sciences Journal*, 56(4): 662–670.

Burton, M. and Molden, D. 2005. Making sound decisions: Information needs for basin water management, in M. Svendsen, ed., *Irrigation and River Basin Management: Options for Governance and Institutions*, Wallingford: CABI, pp. 51–74.

Chowdhury, N.T. 2010. Water management in Bangladesh: An analytical review, *Water Policy*, 12(1): 32–51.

Conca, K., Wu, F. and C. Mei. 2006. Global regime formation or institution building? The principled content of international river agreements, *International Studies Quarterly*, 50(2): 263–285.

Cronin, A.A., Prakash, A., Priya, S. and Coates, S. 2014. Water in India: Situation and prospect, *Water Policy*, 16(3): 425–441.

Dahal, S.H., Gazdar, H., Keethaponcalan, S.I. and Murthy, P. 2003. *Internal Conflict and Regional Security in South Asia: Approaches, Perspectives and Policies*, Geneva: UNIDIR.

Das, S. 2011. Cleaning of the Ganga, *Journal of the Geological Society of India*, 78(2): 124–130.

Dimitrov, R. 2006. Inside Copenhagen: The state of climate governance, *Global Environmental Politics*, 10(2): 18–24.

Fischhendler, I. 2008. Ambiguity in transboundary environmental dispute resolution: The Israel–Jordanian water agreement, *Journal of Peace Research*, 45(1): 91–110.

Gerlak, A.K., Lautze, J. and Giordano, M. 2011. Water resources data and information exchange in transboundary water treaties, *International Environmental Agreements: Politics, Law and Economics*, 11(2): 179–199.

Grumbine, R.E. and Pandit, M.K. 2013. Threats from India's Himalaya dams, *Science*, 339(6115): 36–37.

Grumbine, R.E. and Xu, J. 2011. Mekong hydropower development, *Science*, 332(6026): 178–179.

Gupta, A.K. and Wang, H. 2009. China and India: Greater economic integration, *China Business Review*, 1 September. Available at: www.chinabusinessreview.com/china-and-india-greater-economic-integration/ (accessed 12 July 2017).

Huang, Z. 2015. Case study on the water management of the Yaluzangbu/Brahmaputra River, *Georgetown International Environmental Law Review*, 27(2): 229–250.

Huber, A. and Joshi, D. 2015. Hydropower, anti-politics and the opening of new political spaces in the Eastern Himalayas, *World Development*, 76: 13–25.

Intelligence Community Assessment (ICA). 2012. *Global Water Security*, IC coordinated paper ICA 2012-08. Available online at: http://fas.org/irp/nic/water.pdf (accessed 5 September 2016).

Jia, S., Lv, A., Han, Y., Long, Q., Zhu, W. and Yan, H. 2014. *Water Resources Security Report of China*, Beijing: Science Press.

Jia, S., Sun, Y., Svensson, J. and Mukherjee, M. 2016. Comparative analysis of water rights entitlements in India and China, *Water Policy*, 18(S1): 50–67.

Karkkainen, B.C. 2006. Managing transboundary aquatic ecosystems: Lessons from the Great Lakes, *Pacific McGeorge Global Business & Development Law Journal*, 19(1): 209–240.

Kattelus, M., Kummu, M., Keskinen, M., Salmivaara, A. and Varis, O. 2015. China's southbound transboundary river basins: A case of asymmetry, *Water International*, 40(1): 113–138.

Khalequzzaman, M. 2006. Large dams in NE India: Implications for Bangladesh. Available at: www.learningace.com/doc/5451611/3720cc46ebb9e335386e7feb8a556dd1/dams-in-ne-india-implications-for-bangladesh (accessed 30 August 2017).

Khan, T. 1996. Management and Sharing of the Ganges, *Natural Resources Journal*, 36(3): 455–479.

Li, Z. 2013. Indian threat-mongering over water resource disputes dangerous fantasy, *Global Times*, 7 October. Available at: www.globaltimes.cn/content/815960.shtml (accessed 19 November 2015).

Miles, E.L., Underdal, S., Andresen, S., Wettestad, J., Skjæreth, J.B. and Carlin, E.M., 2002. *Environmental Regime Effectiveness: Confronting Theory With Evidence*, Cambridge, MA: MIT Press.

Mitchell, S.M. and Zawahri, N.A. 2015. The effectiveness of treaty design in addressing water disputes, *Journal of Peace Research*, 52(2): 187–200.

Nandimath, O.V. 2007. *Oxford Handbook of Environmental Decision Making in India: An EIA Model*, Delhi: Oxford University Press India.

Panda, J. 2013. Competing realities in China–India multilateral discourse: Asia's enduring power rivalry, *Journal of Contemporary China*, 22(82): 669–690.

Panda, J. 2015. Future of India–China boundary: Leadership holds the key, *Strategic Analysis*, 39(3): 287–292.

Panda, J.P. 2016. *India–China Relations: Politics of Resource, Identity and Authority in a Multipolar World Order*, Abingdon: Routledge.

Prasai, S. and Surie, M.D. 2015. Water and climate data in the Ganges Basin: Assessing access to information regimes and implications for cooperation on transboundary rivers, *Water Alternatives*, 8(2): 20–35.

Price, G. 2014. *Attitudes to Water in South Asia*, Chatham House. Available at: www.chathamhouse.org/publication/attitudes-water-south-asia (accessed 30 August 2017).

Rahaman, M.M. 2006. The Ganges water conflict: A comparative analysis of 1977 Agreement and 1996 Treaty, *Asteriskos*, 1/2: 195–208.

Rahaman, M.M. 2009. Integrated Ganges basin management: Conflict and hope for regional development, *Water Policy*, 11: 168–190.

Ramachandran, S. 2006. India, Bangladesh fight against the current, *Asia Times*, 8 June. Available at: www.atimes.com/atimes/South_Asia/HF08Df04.html (accessed 12 July 2017).

Sahoo, N. 2014. Decoding Modi's foreign policy, *Carnegie Foundation for International Peace*, 23 September. Available at: http://carnegieendowment.org/2014/09/23/decoding-modi-s-foreign-policy-pub-56711 (accessed 10 July 2017).

Salman, S.M.A. 2010. Downstream riparians can also harm upstream riparians: The concept of foreclosure future uses, *Water International*, 35(4): 350–364.

Seth, B.L. 2013. Li's visit yields little on Brahmaputra, *Down to Earth*, 22 May. Available at: www.downtoearth.org.in/news/lis-visit-yields-little-on-brahmaputra--41141 (accessed 12 July 2017).

Singh, S. 2013. The economics of Sino-India relations [zhongyin guanxi de jingji xue], *Dongfang Daily*, 31 October [in Chinese].

Sood, A. and Mathukumalli, K. 2011. Managing international river basins: Reviewing India-Bangladesh transboundary water issues, *International Journal of River Basin Management*, 9(1): 43–52.

Stinnett, D.M. and Tir, J. 2009. The institutionalization of river treaties, *International Negotiation*, 14(2): 229–251.

Stobdan, P. 2013. BDCA with China and its implications for India, *IDSA Comment*, 29 October. Available at: www.idsa.in/idsacomments/BDCAwithChinaanditsImplications forIndia_pstobdan_291013.html (accessed 12 July 2017).

Sun, J., Jia, S. and Lv, A. 2011. Information extraction and analysis of the Himalayan international rivers [Ximalaya diqu guoji heliu xinxitiqu yu fenxi], *South-to-North Water Transfers and Water Science & Technology*, 9(3): 33–38.

Timmerman, J. 2004. Incorporating user needs into environmental information systems, in J. Timmerman and S. Langaas, eds, *Environmental Information in European Transboundary Water Management*, London: IWA Publishing.

Uitto, J.I. and Duda, A.M. 2002. Management of transboundary water resources: Lessons from international cooperation for conflict prevention, *The Geographical Journal*, 168(4): 365–378.

Verma, R. 2013. The tiger and the dragon: A neoclassical realist perspective of India and China in the oil industry in West Africa. Doctoral dissertation, London School of Economics.

Von Stein, J. 2008. The international law and politics of climate change: Ratification of the United Nations Framework Convention and the Kyoto Protocol. *The Journal of Conflict Resolution*, 52(2): 243–268.

WRG (The 2030 Water Resources Group). 2009. *Charting Our Water Future: Economic Frameworks to Inform Decision-Making*, McKinsey & Co., London. Available at: www.2030waterresourcesgroup.com/water_full/charting_our_water_future_final.pdf (accessed 12 July 2017).

Yan, J. Forthcoming. Assessment of China's transboundary water resources and water use, doctoral thesis, Chinese Academy of Sciences, Beijing.

Zawahri, N.A. 2008. Designing river commissions to implement treaties and manage international rivers, *Water International*, 33(4): 464–474.

Zawahri, N.A. 2009. Third party mediation of international river disputes, *International Negotiation*, 14(2): 281–310.

# 7 Central Asia

## Sino-Kazakh water diplomacy on the Ili and Irtysh Rivers

### Introduction: Sino-Kazakhstan relations

Kazakhstan, a central Asian country, is located in a strategic geographical position bordering China in the north-west. Strong suspicion existed among Chinese policymakers in the early 1990s after Kazakhstan became an independent state, particularly given the uncertain stability of the pre-Soviet states that were experiencing societal upheavals (Xing, 1996). Impacted by its regime transition and differences in ideological beliefs, foreign relations between China and Kazakhstan have developed slowly. After a bumpy start, China has developed relatively stable foreign relations with Kazakhstan. Only since the late 1990s has China begun to develop increasing diplomatic initiatives with Kazakhstan. Such initiatives have borne success in the settlement of border disputes with Kazakhstan.

China and Kazakhstan share 23 rivers, of which the Ili (Yili) and the Irtysh (Erqisi) are the two longest (see Figure 7.1). Both China and Kazakhstan face challenges in coping with water scarcity and their competition over shared river resources is acute, in particular over water quantity. For China, both the Ili and Irtysh Rivers are major sources of water for Xinjiang (see Table 7.1). For Kazakhstan, more than 43 per cent of its water is sourced from abroad. Of this, 43 per cent comes from China, and Uzbekistan, Russia and Kyrgyzstan account for another 33 per cent, 17 per cent and 7 per cent, respectively. China therefore accounts for the highest proportion of water flowing into Kazakhstan.

Kazakhstan disagrees with China's policy over the planning of shared rivers. It is concerned that China may show little consideration about its unilateral use of the Ili and Irtysh Rivers to develop industry in Xinjiang, or to secure the region's water supplies for irrigation, oil exploration and drinking water. Not much different from other cases of water sharing, water quantity constitutes one of the most frequently identified sources of water tensions (Haftendorn, 2000; Yoffe *et al.*, 2003; Tir and Ackerman, 2009). Strong disagreements exist between China and Kazakhstan over how to share the Ili and Irtysh Rivers. Negotiations between the two countries are particularly centred on the division of water quantity, as both countries are experiencing a rapid increase in water demand along with substantial economic development. In particular, both

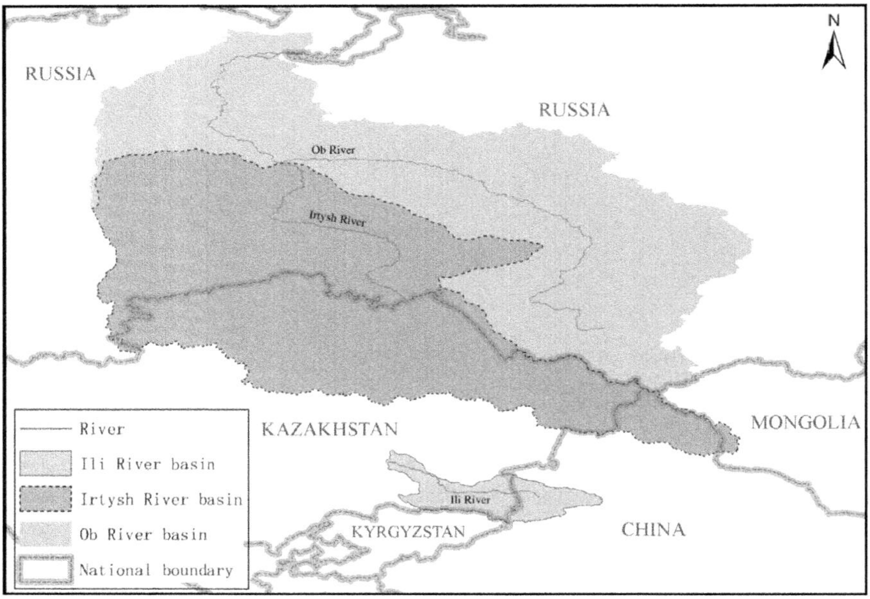


*Figure 7.1* The geographical location of the Ili and Irtysh Rivers.
Source: authors' commissioned work, credit: Jiabao Yan.

countries have displayed strong interdependence on the Ili; this situation represents a challenging situation in which to develop substantive cooperation.

In various areas the two countries have developed diplomatic cooperation, which has laid a solid foundation on which to negotiate on disagreements over shared rivers. Diplomatic cooperation between China and Kazakhstan has also expanded to economic and trade agendas, coinciding with socio-economic improvements in Kazakhstan.[1] In 2006, the country moved into the upper-middle-income group of nations. Natural resource trade accounts for a large proportion of Kazakhstan's economic and trade cooperation with China. Since 2006, when the first Sino-Kazakh oil pipeline was established, Kazakh oil exports to China have grown steadily. In recent years, China has developed a Silk Road Economic Belt strategy that aims to link the continents of Asia and Europe. Xinjiang, China's north-west region adjacent to Kazakhstan, is set to be a gateway for extended business and trade activities with central Asia (Wang and Jia, 2016). Such interest was shared by Kazakhstan, which has been seeking investment opportunities to boost its domestic economy. China's launch of the Silk Road Economic Belt project in 2015 has marked expanded economic and trade cooperation between the two countries. And this move has promoted China to become Kazakhstan's second largest trading partner, with over US$15 billion in trade agreements between the two countries in areas including steel,

*Table 7.1* China's and Kazakhstan's water resource management in the basins of the Ili and Irtysh Rivers

| | | *River catchment area ($km^2$)* | *Runoff depth (mm)* | *Water resources ($10^9 m^3$)* | *Agricultural water withdrawal ($10^9 m^3$)* | *Industrial water withdrawal ($10^9 m^3$)* | *Domestic water withdrawal ($10^9 m^3$)* | *Total withdrawal ($10^9 m^3$)* | *Ratio of China's uses of water resources to the total basin (%)* |
|---|---|---|---|---|---|---|---|---|---|
| Ili River | China | 56,932.43 | 275.54 | 156.8729 | 52.2260 | 1.1744 | 3.3763 | 56.7767 | 36.2 |
| | Kazakhstan | 94,934.23 | 88.76 | 84.2642 | 26.9872 | 0.3738 | 8.5829 | 35.9438 | 42.7 |
| Irtysh River | China | 56,992.3 | 176.79 | 100.7566 | 14.4860 | 0.4567 | 0.5055 | 15.4482 | 15.3 |
| | Kazakhstan | 804,382.04 | 33.28 | 267.7000 | 22.3600 | 1.1943 | 10.8365 | 34.3898 | 12.8 |

Source: Yan (forthcoming).

metals, hydropower generation and a variety of industrial sectors.[2] Encouraged by the national government in Beijing, diversified private actors have emerged resulting in an increasing number of Chinese private corporations investing in more areas of Kazakhstan's economy.

Additionally, China and Kazakhstan have developed a strategic partnership on defence. Border security is the most natural sphere for security cooperation between Kazakhstan and China due to the perceived threat posed by the Uighur population who reside immediately south-east of Kazakhstan. The two countries have also developed and strengthened their cooperation in military defence, signalled by Kazakhstan's joining of the Shanghai Cooperation Organisation (SCO). Both countries have been committed to ensuring border security in their fight against the so-called 'Three Evils' of separatism, extremism and terrorism (Clarke, 2014).

Nevertheless, the sharing of the Ili and the Irtysh Rivers has become a source of strain in foreign relations between the two countries. Since China is not party to the UNECE or the UNWC, Kazakhstan has been concerned that China may not voluntarily perform its water-sharing responsibilities. It is particularly concerned that China shows no intention to consult with the Kazakh government, and that non-water issues such as oil and gas exports may be used as a bargaining chip for China to implement its water policies to secure its domestic economic interests.

This chapter examines the complex nature of water diplomacy developed between China and Kazakhstan over the Ili and Irtysh Rivers. The second part reviews the features of the river basin of the shared rivers in north-west China. In the third part, Kazakhstan's position on water management and its disagreement with China over water-sharing is examined. The fourth part discusses China's diplomatic cooperation before drawing tentative conclusions. It is argued that China's diplomatic initiatives have not yet settled Kazakhstan's disagreements, particularly with regard to the division of water quantity as water scarcity becomes an increasingly pressing domestic issue for both countries. It remains unclear how China can successfully tackle its water disagreements with Kazakhstan without having to make significant changes to its current water policies in Xinjiang.

## River characteristics

### *Water security on the Ili and Irtysh Rivers*

The Ili River flows into the Western Balkhash region, flowing from the Ili Kazakh Autonomous Prefecture of the Xinjiang Uighur Autonomous Region to the Almaty Province in Kazakhstan where it terminates in Lake Balkhash. Both Xinjiang Province in China and Kazakhstan have shown high demand of water for agricultural activities. For instance, water from the Ili and Irtysh Rivers is mainly used for irrigation in Kazakhstan.[3] Agriculture forms less than 10 per cent of the nation's GDP, among which cotton accounts for a large proportion of the total income from agriculture. Kazakhstan's production ranks it the world's top eight grain-producing countries. It is predicted that investment aimed at increasing the productivity of grains is likely to occur, requiring further water for irrigation (Granit *et al.*, 2010). Insufficient water supply for irrigation has been regarded as a key concern for maintaining stability in the country's rural areas.[4]

Moreover, China and Kazakhstan have shown heavy reliance on the Ili River. With regard to their exploitation of water resources in the Ili River basin, both countries have displayed a high rate, reaching 36.2 per cent and 42.7 per cent for China and Kazakhstan respectively.[5] Yan (forthcoming) indicates that the Ili River has started to show signs of declining water security. It is predicted that the Ili River basin may soon fail to provide secured water resources for both China and Kazakhstan.[6] Hence, both China and Kazakhstan are experiencing serious water insecurity on the Ili River.

The Irtysh River originates in China on the western slopes of the Mongolian Altai Mountains and is called the Black Irtysh until it flows into Lake Zaysan in Kazakhstan. It flows through Kazakhstan and onwards into the Ob River basin in Russia, constituting an important water source for the Omsk Oblast of southern Siberia, as well as for northern Kazakhstan. China and Kazakhstan have also shown intensified exploitation of the Irtysh River. In particular, Kazakhstan has shown an alarming concern over water scarcity for agricultural and economic growth, and the ecological degradation of the rivers it shares with China. Its reliance on the Ili River's water resources has increased to an unsecured level (see Table 7.2). The Irtysh River also contributes heavily to Kazakhstan's domestic consumption. The Irtysh River is the main source of water for around four million residents of northern Kazakhstan and three of its large industrial cities, namely Karaganda, Semipalatinsk and Pavlodar (Blua, 2004). Apart from direct competition in water extraction, the two countries have also shown a growing rate of use of the water resources of the river basin. China possesses less than one-third of the water resources in the Irtysh River

*Table 7.2* Features of the Ili and Irtysh Rivers

| | *Ili River* | *Irtysh River* |
|---|---|---|
| Length (km) | 1,236.5 | 4,248 |
| Length (within China) (km) | 442 | 633 |
| Length within Kazakhstan (km) | 793.5 | 1,698 |
| Length within Russia (km) | 0 | 1,964 |
| Basin areas | 1,512,000 | 1,640,000 |
| Basin areas within China ($km^2$) | 56,932.4 | 56,992.3 |
| Percentage to Xinjiang's total runoff water (%) | 18.6 | 13.5 |
| Annual runoff within China (billion cubic metres) | 16.44 | 11.1 |
| Ratio of water use to water resources in China | 36.2% | 23.5% |
| Ratio of water use to water resources in Kazakhstan | 42.7% | 12.8% |

Sources: Tang and Li (1999); Jia and Liu (2014).

basin, whereas Kazakhstan possesses more than double China's. However, with a higher level of water extraction, China's water resource exploitation rate almost doubles that of Kazakhstan's (see Table 7.1).

### *China's water security and Xinjiang policy*

Both the Ili and Irtysh Rivers are major sources of water for Xinjiang (see Table 7.2). China is faced with mounting pressure to quench Xinjiang's demand for water and the declining level of water security in Xinjiang has been exacerbated by characteristics of the Ili and Irtysh Rivers. In particular, Xinjiang's water resources are unevenly distributed, physically. At a temporal scale, runoff from the Ili River mostly occurs within a four-month period between June and September with the highest level appearing in July (Tang and Li, 1999). In comparison, the Irtysh River has a higher level of accumulated runoff between May and August with the highest level appearing in June. In terms of spatial distribution, the Ili and Irtysh Rivers are located in the western part of Xinjiang, leaving the eastern part of the province with fewer tributaries. Despite the basin area of the two rivers covering only 10 per cent of Xinjiang, they provide a third of the province's total water resources. Therefore, Xinjiang Province's reliance on the Ili River to meet this demand has reached an alarming level.

There has been an increase in exploration of the Ili and Irtysh Rivers over the past decade, as their river basins are believed to form one of the largest areas whose potential has not yet been explored (Zhou and Zhang, 2005). As a result of the Chinese national government's 'Go West' initiative, Xinjiang has developed a fast-growing population that includes substantial numbers of people relocating from other provinces. Xinjiang's population has increased to 23.6 million (see Figure 7.2), up from only 16 million as recently as the mid-1990s (Stone, 2012). The Beijing government has planned to develop a series of new towns and urban areas that surround the city of Wulumuqi (see Figure 7.3).[7] Such a fast-growing population in the western part of the country is putting increasing areas of land under cultivation and irrigation.

The spatial distribution of water resources has also appeared to be a constraint to industrial development in Xinjiang. China's 2003 white paper on Xinjiang has made a case for the expansion of energy industries as part of the ongoing 'Go West' initiative. However, the water reserves in the planned Tulufan and Hami industrial zone amount to only 2.2 per cent of Xinjian's usable water resources (Li and Li, 2014). The Tianshan north industrial zone, which is another part of the region planned for major development, generates over half of the province's GDP revenue but has access to only 7.4 per cent of the province's overall water resources. Moreover, Beijing has also been active in exploiting Xinjian's oil resources, with a focus on the fields in the Tarim Basin which are rich in oil reserves and are estimated to amount to 30 per cent of the overall reserves in the country.[8] For instance, the Tulufan field, one of three large fields in the Tarim Basin, is estimated to hold ten billion tonnes of oil. However, these fields are situated in the desert where inhabited areas are few

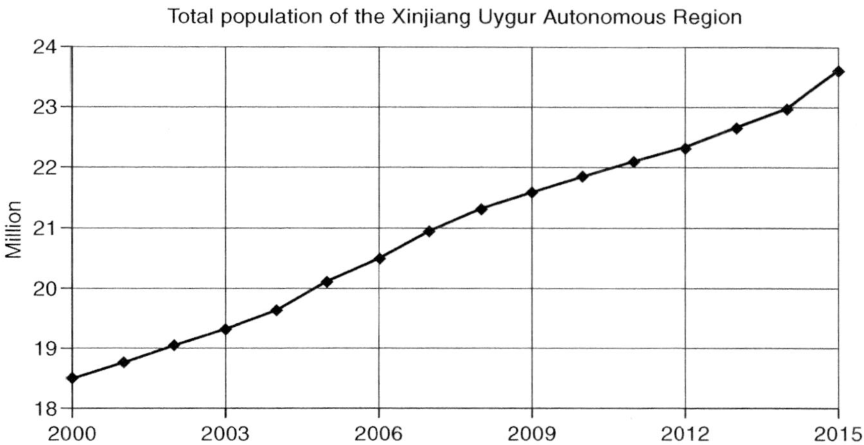


*Figure 7.2* Xinjiang's population growth from 2000 to 2015.

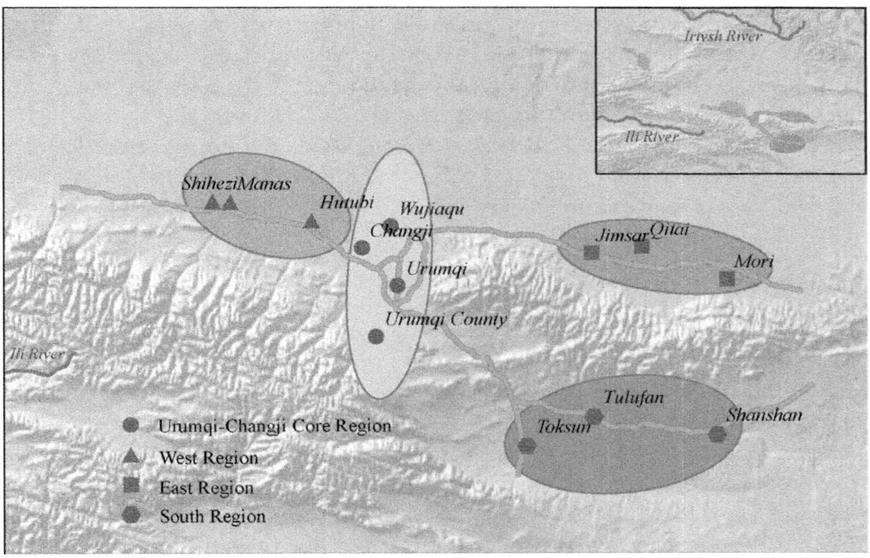


*Figure 7.3* The development of clusters of Wulumuqi city cluster.

Source: authors' commissioned work, credit: Jiabao Yan.

and far between and their exploitation demands water resources (Deng, 2010). Such a situation has alerted the Beijing government in its planning and management of Xinjiang's water resources.

To resolve the issue of water scarcity for industrial development, Beijing has developed an extensive water storage and irrigation network. Since the 1990s, a '635 water diversion project' has been launched but not yet completed at the time of writing. It aims not only to supply the Irtysh–Karamay–Urumqi Canal with water, but also to reserve water for this region. By 2016, construction of its 324 km-long canal to reroute water from the Irtysh River to the north piedmont of Tianshan Mountain was completed. And 0.43 billion $m^3$ of water has been reserved. The canal transfers water from the Irtysh River further south to Karamay (Kelamayi) and Wulumuqi, where it is used for both irrigation and for the general use of the population and industries. This project will also transfer water to East Xinjiang for irrigation and ecological uses and to support the oil industry.

Another factor in the water scarcity issue is Xinjian's low efficiency in water use. It is estimated that industrial water use in the province has shown a much lower efficiency rate than other cities in China. It is noted that agricultural water use has accounted for over 95 per cent of Xinjiang's overall water use, and this has seriously impeded the management of water resources (Deng, 2010). With regard to the Irtysh River specifically, a new irrigation project aimed at improving the water situation has been developed. This is a key project that promotes the agricultural and industrial foundations of Western China. Through water planning, for instance, some of the river basins in southern Xinjiang have been incorporated into the government's plan for development. These include the southern Xinjiang, the Hetian (Hotan), the Yarkand, the Kashgar, the Aksu and the Korla oases. Among these, the Junggar Basin, which is located between the Tian and Altai Mountains, is regarded as another important water source in the region (Chen, 2014).

## Kazakhstan's positions on water-sharing

### *Kazakhstan's domestic water management*

Kazakhstan has realized certain economic growth, and to continue promoting its economic development and citizens' well-being Kazakhstan has been placing a greater strategic emphasis on the management of its water resources. In both water policy and foreign policy development, multiple interests have been identified to promote sustainable development, which incorporates multiple areas of improved water management. Indeed, water has been regarded as a necessary source for Kazakhstan's national development. Since the early 2000s, this has been repeatedly confirmed by Kazakh government's policy documents (see Table 7.3). This section reviews Kazakhstan's areas of concern that are in dispute with China over the management of transboundary rivers.

The quality of water and the prevention of river ecosystem degradation are key issues for Kazakhstan's water management, as demonstrated by several

*Table 7.3* Water management emphasized in key policy documents

| | *Clean water and water quality control* | *Quantity* | *Conservation of water ecology* | *Water resource as a key part in sustainability* |
|---|---|---|---|---|
| The Kazakhstan Water Code | ✓ | ✓ | ✓ | ✓ |
| The state programme on industrial development of Kazakhstan for 2015–2019[1] | | | ✓ | ✓ |
| Concept on transfer to Sustainable Development 2007–2024 | ✓ | | ✓ | ✓ |
| Kazakhstan-2030 Strategy for development[2] | ✓ | ✓ | | |
| The Strategy Kazakhstan-2030 | ✓ | ✓ | | ✓ |

Notes:

1 https://strategy2050.kz/static/files/pr/eng.docx (accessed 4 November 2016). The water code is a founding legal document that sets out guidance for national policy on water management.

2 http://prokuror.gov.kz/eng/state/acts-president/strategy-kazakhstan-2030 (accessed 4 November 2016).

national policy documents including the mid-term strategy (2011–2050).[9] The water management targets outlined in this strategy also aim to guarantee access to clean water. To secure the water supply, institutional reforms have been undertaken to ensure a high level of self-sufficiency in agricultural production. Furthermore, saving water has been prioritized as key area for technological improvements to be made. Improving the technologies in this regard has been identified as a solution, and is therefore a key target for advancing the country's industrialization.

Conservation of water ecology is another important issue highlighted in Kazakhstan's national policies. The Kazakh government recognizes that the sustainable use of water resources and the prevention of water pollution are key targets to realizing environmental stability, as prescribed by Article 144(1) of the nation's Water Code.[10] This chapter discusses the effective and rational use of transboundary waters for maintaining and rehabilitating the biological balance in and surrounding these bodies of water. It recognizes that in national economic development, sustainably profiting from water resources must be balanced with the maintenance of the biosphere.

Kazakhstan has concerns about the negative impacts on water ecology that result from China's exploitation of natural resources in Xinjiang. Such concerns are also shared by Chinese scientists (Tai *et al.*, 2003). Domestic and industrial discharges into both the Ili and Irtysh Rivers in China have resulted in increased pollution and ecological damage to these rivers (Zhou *et al.*, 2003; Li and Li, 2014). It is believed that Xinjiang's ecological conditions are already fragile, and such pollution thereby greatly threatens the ecology of these rivers (Tai *et al.*, 2003). Additionally, scientists from both countries also fear that Lake Balkhash in Kazakhstan, which is fed by the Ili River, will be polluted with chemicals and fertilizers as a result of the rapid industrial development of northwestern China. Furthermore, it has been estimated that about 8,000 small lakes along the Ili River have recently dried up because of the overexploitation of water resources in the region. Additionally, with regard to Lake Balkhash in particular, a substantially reduced water inflow has led to growing concerns about the potentially rapid diminishment and desertification of the lake.

### *Kazakhstan's positions and foreign policy on water*

Kazakhstan's foreign policy has been largely determined by features of its water management. These are seen in the country's adopting global norms and developing basin-wide governance as preferred mechanism of water management institutions (WMIs). However, strong disagreements exist between the Kazakh and Chinese governments' interpretation of global water norms.

In foreign policy, water is generally considered a strategic interest in Kazakhstan's political relations with other nations. As the country is located downstream on rivers that are key water sources, Kazakhstan has realized that cooperation with upstream countries is an absolute necessity to promote sustainable water management, and it is thus a founding principle for Kazakhstan in the sharing of international rivers. In 2000, Kazakhstan became a member of the UNECE Convention, being among this convention's earliest subscribers. This indicates that it has adopted the UNECE's principles for water-sharing and resolving the problems that arise in such processes. Since the early 2010s, in its diplomatic practices, Kazakhstan has been proactive in establishing institutional cooperation with other nations to share the use of transboundary water resources. In its foreign policy paper (2014),[11] promoting cooperation on transboundary water is given as a strategic interest. Essentially, by so doing, Kazakhstan's government has indicated that the country is willing to commit to international cooperation to resolve its water insecurity.

Kazakhstan and China contrast in their positions on how to best exercise their sovereign rights to appropriate their water resources fairly. These differences have also become sources of disagreements. For instance, Kazakhstan's approach has incorporated cooperation and a 'no harm' principle, the government advocating that responsibilities are taken into consideration in national policymaking relating to water uses by other nations. In particular, the government has concerns about water uses that may have a 'transboundary impact'. In its Water Code, Kazakhstan applies the principle of 'no harm' by monitoring activities and sharing information. For instance, according to Article 143(8) of the Water Code, countries are obligated to monitor water information and avoid water emergency situations or disasters by the timely exchange of such information with other nations. Furthermore, to uphold its consideration of fairness in the management of its shared water resources, Kazakhstan has also adopted environmental norms such as 'polluter pays', the 'precautionary principle' and others for water cooperation. Its Water Code indicates that economic compensation is likely to be considered for transboundary river issues when necessary, with Article 144(1) and (2) further indicating that it is willing to adopt economic measures to support the principle of 'equity participation' and the fair allocation of shared waters. From past experiences, Kazakhstan has agreed to provide financial or material compensation for water management services provided, as has been seen in the agreement signed with Kyrgyzstan, to which Kazakhstan supplies natural gas and oil. It can be inferred that Kazakhstan is willing to further adopt this principle in its cooperation on transboundary water management with other neighbouring countries (Vogtmann and Dobretsov, 2003; UNECE, 2009).

Kazakhstan strongly advocates for basin-wide governance, which rather differs from China's preference for bilateral negotiations. Kazakhstan has supported the norm to incorporate all relevant riparian states in river basin management for sustainable development. To Kazakhstan, the issue of water shortages is threatening the stability and security of the region, and it is thought that this issue would be best addressed by developing inclusive basin-wide cooperation. This requires that all parties with an interest in a given water basin are regarded as important actors. It advocates that the extent of management is based on an understanding of the issue of scale, such as that of river basins. Kazakhstan believes that governance should be arranged so as to include various actors from beyond its national borders. Kazakhstan's cooperative efforts with other riparian states have seen the successful establishment of institutional cooperation in various areas of sustainable water management including water-saving measures and the prevention of disasters and emergencies. For instance, Kazakhstan has established a joint water management committee with Russia and Kyrgyzstan,[12] and is proactively seeking broader cooperation with all five of the central Asian countries that formerly relied on Soviet institutions to organize them, namely Russia, Tajikistan, Kyrgyzstan, Turkmenistan and Uzbekistan.[13]

However, with regard to the preferences for water-management institutions, China has also shown an interest to persist with bilateral cooperation with Kazakhstan. Its position was hardly affected when Kazakhstan made an attempt for multilateral cooperation. With an intention to exert pressure on China, Kazakhstan has advocated trilateral cooperation over the sharing of the Ili and Irtysh Rivers.[14] Regarding cooperation over the Ili River that runs through both countries, Kazakhstan insists on incorporating Kyrgyzstan in any talks because the river also runs through that country. Efforts to facilitate this trilateral cooperation have also been made through the SCO, a multilateral forum ostensibly designed to promote cooperation across a range of areas. As China is also a member of the SCO,[15] Kazakhstan has attempted to leverage with China and make improved progress on addressing the two nations' water-sharing issues.[16] Similarly, regarding the Irtysh River, Kazakhstan insists that Russia should join efforts to reach a comprehensive agreement with China. However, there is no indication of transboundary water management issues ever being raised within the SCO framework as a topic for discussion with both Kazakhstan and Kyrgyzstan. Although Kazakhstan was keen to pressure China to negotiate on the issue of water volume, progress on this key concern has developed rather slowly. China and Kazakhstan have not yet established common ground on what a fair water division is and how it can be reached.

## China's water diplomacy

Along with rapid economic development in Xinjiang, China has established closer interaction with Kazakhstan (Adeel and Wirsing, 2016). With regard to water diplomacy, cooperation between the two countries has not been smooth. Although cooperation began from the late 1990s, it was not until after 2006 that China began to attach a heightened significance to the issue of sharing water with Kazakhstan. It is noted that both countries' collaboration in water management has expanded from sporadic concerns to a much broader scale and focused level of cooperation (see Table 7.4).

In the early stages of Kazakhstan–China cooperation, China focused on exercising its unilateral interests in water-resource management. China insisted it would maintain its sovereign rights to appropriate its water resources. In the words of the government spokesman Jianchao Liu, 'on the issue of cross-border rivers, the Chinese government will continue to adopt a responsible attitude,

*Table 7.4* Agreements formed between Kazakhstan and China on the management of shared rivers

| *Date* | *Agreement* |
|---|---|
| 1995 | Agreement on Cooperation in Meteorological Technology between China's Ministry of Water Resources and Kazakhstan's national Meteorological Administration |
| 2001 | Agreement between the governments of Kazakhstan and China for cooperation in the use and protection of transboundary rivers (Astana) |
| 2002 | Agreement on water distribution and use of the Khorgos River (Yining) |
| 2004 | Agreement between the local authorities of Almaty Oblast of Kazakhstan and Ili-Kazakh Autonomous Region of China on cooperation in the use of the Sumbe and Kaishibulak (China) transboundary rivers |
| 2005 | Agreement between the Ministry of Agriculture of Kazakhstan[1] and the Ministry of Water Resources of China on emergency notification of the parties about natural disasters on rivers (Astana) |
| 2006 | Agreement between the Ministry of Agriculture of Kazakhstan and the Ministry of Water Resources on the development of research cooperation on transboundary rivers (Beijing) |
| 2006 | Agreement between the Ministry of Environment of Kazakhstan and the Ministry of Water Resources of China on mutual exchange of hydrological and hydro-chemical data of the hydrological stations on the main transboundary rivers (Beijing) |
| 2010 | The cooperation Agreement on the Joint Construction of Horgos River Friendship Joint Diversion Control Project of 2011 |
| 2011 | Agreement between the governments of Kazakhstan and China on the protection of water quality of transboundary rivers (Beijing) |
| 2013 | Agreement between the governments of Kazakhstan and China on the management and operation of the joint combined Dostyk waterworks on the Khorgos River (Astana) |
| 2014 | The cooperation between the Republic of Kazakhstan and the People's Republic of China in the field of use and protection of transboundary water resources |

Note

1 Institutional reforms have taken place to facilitate the implementation of the national water management policy. The management of water resources had been an area within the Ministry of Agriculture's portfolio. In 2013, the Ministry of Environmental Protection incorporated more responsibilities regarding water management into its remit and was transformed into the Ministry of Environment and Water Resources. However, in 2014, Kazakhstan's Ministry of Environmental Protection and Water Resources was closed, and its functions were divided between two ministries, including a newly created Ministry of Energy and the Ministry of Agriculture. Since then, responsibilities for water have been placed back with the Ministry of Agriculture under which a committee for water resources was established.

not harm the interest of our neighbouring countries and properly handle relevant issues through dialogue and consultation'.[17] Nevertheless, China's defence had also been replaced by its commitment to develop more trusting relations with Kazakhstan. China is well aware that, located on the upstream, the country has posed a threat to downstream Kazakhstan.

China has shown a relatively strong commitment to build trust. Closer Sino-Kazakh collaboration on water-sharing was signalled by the signing of the 2001

Agreement on the Protection of Water Quality of Transboundary Rivers. This agreement laid the foundation for the principles of equity and fairness in the two countries' sharing of rivers. For instance, by 'equitable' use, it is agreed that a country is allowed to plan and manage a river domestically without the intervention of the other. Critics suggest that this allows China's unilateral use of water resources to go unchecked, without any further discussion of 'monitoring' nor any agreement reached on the volume of water intake by China. The 2001 agreement, despite establishing principles for collaboration between the two countries, has therefore been regarded by some as a 'one-sided' agreement that favours China (European Dialogue, 2011).

How to define a 'fair' division of water has become a central question. Fearing that fruitless negotiations would impede wider foreign relations, China has begun to show a more responsive attitude towards negotiations. It has focused on the issue of dividing water by seeking the 'rational utilization' of cross-border water resources.[18] China's position has changed from developing selective diplomacy on water management to an expanded collaboration that incorporates some of the key concerns Kazakhstan has expressed about sharing water between the two nations. For instance, the Chinese negotiators have developed their own principles on shared water resources. China's understanding of fairness focuses on country's closeness and ability to access water resources. As mentioned in Chapter 2, the Chinese side interprets equity as a right to be enjoyed by those who have gained an early access to water resources. Under this principle, the Chinese government has developed its own positions on how to divide water with Kazakhstan. These approaches are based on how to best manage abstractions of its water resources. Such ambiguous approaches have raised questions in debates among experts, policymakers as well as legal practitioners. Similarly, the application of this principle has also been criticized by the Kazakh government.

The Chinese government also takes the position that a country's rights to water are determined by the size of the population that needs to access water resources. This approach would provide a distinct advantage for the country over a 50:50 division (Stone, 2012). As China possesses the greater population, naturally, such a rationale has been immediately rejected by Kazakhstan, given it would result in the nation being allocated a smaller proportion of water than China (Tengrinews, 2013b; Tursun, 2013).

At the same time, it is also believed that timing is an important criterion for water users in their appropriation of water resources. Those who first appropriate water are granted the rights to that water. From a historical perspective, China deems that Xinjiang has used a lower volume of water than its Kazakhstani counterparts. To the Chinese, the late start has permitted Xinjiang the right to use more water at the current time (Zhang, 2015). This point aims to bring support to defend China's hastened policies promoting the development of Xinjiang, although dialogue on shared water uses has been ongoing (Tengrinews, 2013a).

As acute competition exists over water volume, Kazakhstan feels it has made adjustments to accommodate China's preferences and that compromises have been made by maintaining bilateral mechanisms for resolving the water sharing issue (Baizakova, 2015).

### *Areas of collaborative water management*

Despite disagreements about competition over water quantity, both Kazakhstan and China have closely followed the agreed institutional cooperation. Since 2001, more than ten joint committee meetings and separate working group meetings have been held. Under the leadership of the Joint Committee on Cross-Border River Utilization and Protection, coordinated activities have been developed that elevated the committee into a so-called China–Kazakhstan Cooperation Commission in 2008, which functions at a vice-premier level.

China has committed to unilaterally providing Kazakhstan with water data from the two countries' major shared rivers all-year round. Such a move serves to inform Kazakhstan of the China's appropriation of water resources from particular transboundary rivers, thereby reducing suspicion that China may be retaining water data that was important for Kazakhstan to prevent water disasters. By so doing, it was made clear that China had no intention to use water management as a weapon against Kazakhstan, which had limited access to upstream water data.

Among areas of common interest, the control of industrial pollution and disaster prevention have been the focus of collaborative efforts. The primary concern is to develop pollution prevention and management systems for responding to emergencies that could result from industrial discharges. It has been agreed that information sharing constitutes a major principle in collaboration (Article 6 of the Water Code) and that the Kazakhstan Ministry of Environment and the Chinese Ministry of Water Resources will exchange information and provide early warnings about accidental pollution. Such collaborative monitoring has also led to the establishment of a Sino-Kazakh joint commission on the protection of the environment, which includes two working groups focused on the monitoring, analysis and assessment of water quality in the country's transboundary rivers. In particular, the joint commission is made up of representatives from China's Ministry of Water Resources and Ministry of Foreign Affairs, while specific experts and hydrologists from different ministries constitute the membership of the working groups. In 2008, the level of formality was escalated to ministerial-level cooperation, indicating strengthened commitment from both governments.

There have been successful negotiations regarding the prevention of natural disasters, with emphasis placed on flood control. Institutionalized cooperation has been formed. This includes three agreements (signed in 2001, 2006 and 2011) that have been significant in guiding coordinated water management between the two countries. It has been agreed that all-year round hydrological data for the Ili River is to be unilaterally provided by both China and Kazakhstan. On water disaster prevention, collaboration has also occurred in a more flexible fashion. In July 2014, for instance, Kazakhstan experienced droughts which resulted in it calling for assistance from China. The Chinese government promptly responded by allocating additional water for emergency use. According to an official report,[19] for ten days at the Sandaohezi station on the Ili River an extra water volume of 215 $m^3$ per second was released. This extra volume of released water amounted to 50 per cent of the average annual runoff from China to Kazakhstan. With such a commitment, China has agreed to provide more data than its counterpart. In fact, as confirmed in the 2014 declaration made between the two countries, the scope of the information exchanged has been extended to include all of the major shared rivers, including the Irtysh, Ili and Tekes Rivers.[20] This degree of data provision is rarely seen when compared to other situations where China shares international rivers with its riparian neighbours.

Kazakhstan and China have also begun to cooperate on the protection of biodiversity in their shared rivers. The fragile ecological conditions of the shared river basins have been recognized, as have the ecological threats posed by overfishing, industrial pollution and excessive agricultural land uses (Zhou *et al.*, 2003). As early as 2000, a fish rescue centre was established by the Chinese Ministry of Agriculture on the Irtysh River with the aims of protecting the river's many rare species of fish.[21] In 2009 and 2012, the Kazakh National Research Institute of Aquatic Products and the Chinese Bureau of Aquatic Products from Aletai city (Xinjiang) conducted collaborative scientific research

**Box 7.1 The Khorgos project as an integrative water management initiative**

Khorgos is a location where both Kazakhstan and China have concerns about the security of water resources. The Khorgos River, 148 km in length, rises in the Aktas Mountains, an offshoot of the Tianshan Mountains (see Figure 7.4). It flows southwards from Kazakhstan, forming a boundary between the two countries with north-west China's Xinjiang Uighur Autonomous Region on its east bank and Kazakhstan on the west. Both countries face similar water management issues for the Khorgos River, including irrigation and flood control. As such, in 2011 they agreed to jointly construct a structure on the Khorgos River which will allow them to equally share the rivers' waters for irrigation projects. More than 16,000 ha of irrigated land from both countries was covered by the project, based on an accurate accounting of water withdrawal and water supply. The project is also of significance for trade, which demonstrates strategic collaboration in promoting trade and economic activities between the two countries. Transport navigation to the Khorgos–Eastern Gate Special Economic Zone has also benefited from the project (Liang, 2011). Further collaboration has also been made when a joint hydroelectricity system, the Dostyk, was built on the Khorgos with the aim of producing 15 per cent of the total power generated in Xinjiang. Based on the success of the Khorgos River project, it has become a pilot programme that demonstrates how benefit sharing could ease concerns over shared waters.

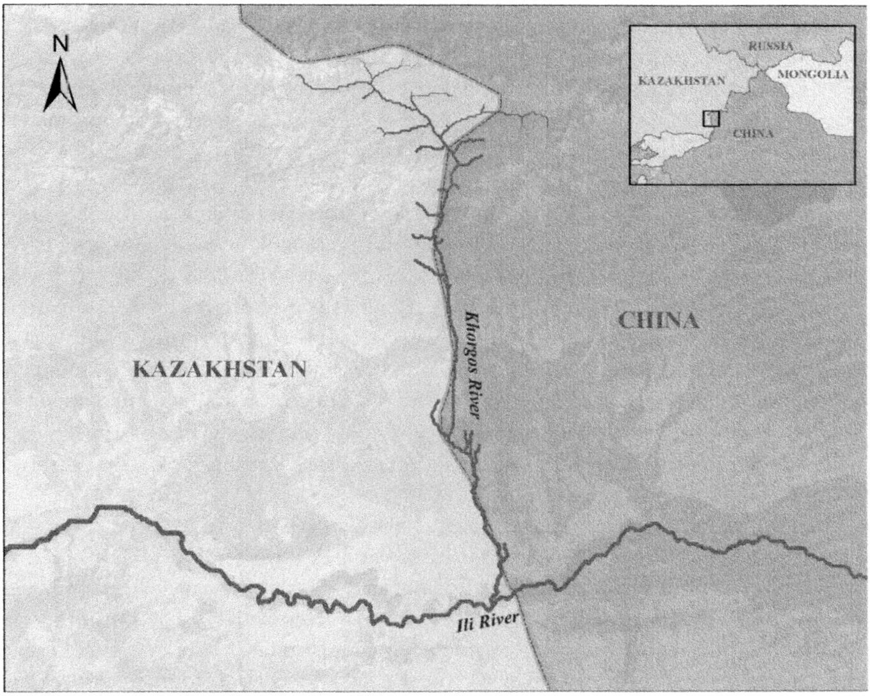


*Figure 7.4* The Khorgos River and its location across national borders.

Source: authors' commissioned work, credit: Jiabao Yan.

to protect hucho taimen and pike perch in the Irtysh River. In both cases, Kazakhstan introduced the fish eggs and China facilitated the incubation with the research aiming to understand the conditions for reproduction at different development stages.[22] However, such collaboration occurred on an ad hoc basis, and this has become more frequent than in the past.

Pragmatism and convenience have been key in the two countries' respective diplomatic activities. Along with the effective implementation of the signed cooperative agreements China has shown an interest in further exploring opportunities to coordinate water management with Kazakhstan. The scope of the joint committee's activities has recently been extended to address a key issue of dispute – water division between the two countries (UNDP, 2011). China, facing increasing pressure from Kazakhstan, has made a gesture to negotiate on downstream water allocations. With an intention to ease tensions between the two sides, China facilitated the development of the Khorgos River sharing project (see Box 7.1). This project serves to benefit water-sharing between the two parties.

***Assessment of water diplomacy***

China and Kazakhstan have managed to peacefully negotiate the management of their shared transboundary rivers. This has largely been the result of increasing pressure from Kazakhstan in response to the country's growing concerns about water insecurity. Despite the development of a number of diplomatic institutions, the two countries have not yet progressed to a point where the shared rivers are managed in a sustainable fashion.

Since the early 2000s, China has become more responsive to Kazakhstan's concerns over the two countries' shared rivers. Water norms have also been strictly followed, albeit in limited areas of water pollution and disaster control. Since 2006 in particular, substantial progress has been achieved through intensified water diplomacy between the two countries. Furthermore, compliance with water-related agreements has been successful, with the establishment of joint committees and working groups. In fact, the constant meetings between experts and ministers has enhanced each side's understanding of the other's needs and concerns, and this has also facilitated the development of substantial water monitoring provisions (UNDP, 2011).

Despite positive diplomatic progress, both Kazakhstan and China have displayed relatively high levels of water insecurity. A key concern for both countries is the issue of water quantity allocations. Negotiations have been held on the basis of the key principles discussed, and have considered the potential for developing collaborative water management initiatives. A breakthrough has been made with the Khorgos River project that allows both countries to realize their interests in water's economic value, while also protecting water ecology. Although the project is rather small in scale it indicates that common issues could potentially be used as the links for promoting collaborative water management. It signals a possible solution to resolving the water quantity issue as a key water sharing concern for both Kazakhstan and China, and may facilitate more substantial collaboration in the future.

## Concluding remarks

Faced with pressure from falling water security, China and Kazakhstan are increasingly aware that water-sharing is a key transnational issue for the future management of the shared rivers that cross their territories. Discussions over how to share water began as early as 1994, two years after diplomatic relations were established, with the larger transboundary rivers, the Ili and the Irtysh being of particular interest. However, there have been troubled negotiations between China and Kazakhstan.

With limited space for negotiation, China's water diplomacy has not been smooth. Despite that both countries have developed close links in various areas of strategic interest, negotiations have been restricted to individual areas of water management. China has shown a commitment to cooperate in peacemaking activities, such as in flood control and drought relief. By so doing, it shows a reserved interest in promoting sustainability in the shared rivers and river basins. China's intention primarily lies in trust-building while proceeding with national interests over the shared rivers. China has committed to maintaining its position on the core dispute of water volume, without showing any intention to compromise on providing more water to Kazakhstan than is deemed necessary. With its commitment to providing water information, China has shown only a little political interest in the management of water data. Nevertheless, with limited collaboration developed in identified areas of water management, China's water diplomacy has had only limited success.

The key disagreement of the division of water quantity has barely been touched upon by the two countries. China has developed principles to highlight its sovereign rights in appropriating shared water resources. China has applied principles developed within domestic water management practices, which are feared to have contributed to an increasing number of disputes in the management of transboundary rivers (Wang *et al.*, 2007). Similarly, with an ambiguous interpretation of the 'equity' principle, China as the upstream country is pushing the boundaries in fulfilling its obligations towards Kazakhstan. Determined by a domestic water management focus, China has developed its own agenda for managing the Ili and Irtysh Rivers. In particular, it has embedded economic and strategic interests that centre around Xinjiang's development when sharing water resources with Kazakhstan.

Kazakhstan has illustrated different interests and principles in its water cooperation with China, and shown a more pronounced insecurity over the shared rivers. In water management, the Kazakh government has proposed a system of sustainable river basin management that considers both water resource utilization and ecological protection. Water scarcity and access to clean water have become key issues in the country's foreign policy agenda, as well as in internal politics. This differs greatly from China's policy environment in that it provides systematic pathways to realizing sustainability over shared rivers, while China's primary interests lie with water resource exploitation. Similarly, China disagrees with Kazakhstan's proposal to build basin-wide cooperation that incorporates all relevant riparian states in negotiations for the management of transboundary rivers. Hence, the two countries have contrasting water-sharing concerns, principles, policies and preferences for management mechanisms. This has resulted in deadlocks in negotiations, despite China's water diplomacy prioritizing human security in neighbouring states.

Another approach that China has adopted in its water diplomacy is to build issue links with other foreign policies. China and Kazakhstan's economic and trade cooperation has developed greatly. As such, it has been suggested that various policy issues have been identified as strategic interests for China regarding Kazakhstan. For instance, Kazakh oil and gas exports are arguably, to a certain degree, a reason behind China's willingness to cooperate on water issues (Biba, 2014). Similarly, it can be said that Kazakhstan's interests in restructuring its economy and enhancing its defence strategy makes it keen to continue negotiating with China along the lines that China has set out on the sharing of transboundary rivers. It recognizes that China has delayed in making a definite agreement on water volume division, and there is great uncertainty about how the country would act responsibly to reduce harms towards Kazakhstan. This may explain why Kazakhstan has shown a pragmatic attitude towards water diplomacy. Until June 2017, when the sides met, they had not reached agreement on the drafting of a water allocation agreement that was set to terminate in 2015. However, Kazakhstan anticipates more economic dividends while also thinking about the possible costs behind Chinese 'help'. With closer relations developed in the areas of trade and the economy, the urgency placed on transboundary water sharing seems to have weakened.

Overall, despite China making successful attempts to address Kazakhstan's concerns over the two countries' shared water resources, such initiatives are far from sufficient to quench Kazakhstan's demands for water. As Kazakhstan is faced with declining water security, it has pressured China to change its position and show more commitment to respecting Kazakhstan's concerns. However, with China's vested interests in water management and regional development, it remains unclear how it can successfully tackle its water disagreements with Kazakhstan without having to make significant changes to its current water policies in Xinjiang.

## Notes

1 World Bank, Kazakhstan Overview: since 2002, Kazakhstan's GDP per capita has risen six-fold and the incidence of poverty has fallen sharply. Available at: www.worldbank.org/en/country/kazakhstan/overview (accessed 25 July 2017).
2 Enhance Sino-Kazakh development and strategy, collectively build partners of common fate and interests, Ministry of Foreign Affairs of the People's Republic of China, 2 April 2015. Available at: www.mfa.gov.cn/web/dszlsjt_673036/t1251460.shtml (accessed 10 April 2017).
3 www.fao.org/nr/water/aquastat/countries_regions/kaz/index.stm (accessed 10 April 2017).
4 See Kazakh national policy document, The Strategy for development of the Republic of Kazakhstan until the year 2030, www.akorda.kz/en/official_documents/strategies_and_programs (accessed 4 November 2016).
5 Water resource exploitation rate refers to the ratio between a country's water extraction and the total water resources.
6 According to global norms, river basins show an alerting level when the ratio of water uses reach 40 per cent.
7 The Wulumuqi urban cluster include the cities of Shihezi, Changji and Wujiaqu.
8 Liu, M. Oil and gas policy reform strategy is displayed in Xinjiang, *Shanghai Securities News*, 23 July 2015. Available at: www.stcn.com/2015/0723/12372716.shtml (accessed 10 April 2017).
9 UNDP Kazakhstan, National Integrated Water Resources Management and Water Efficiency Plan for Kazakhstan. Available at: waterwiki.net/index.php?title=National_IWRM_and_Water_Efficiency_Plan_for_Kazakhstan (accessed 26 July 2017).
10 The full English text of The Water Code of the Republic of Kazakhstan is available at: http://adilet.zan.kz/eng/docs/K030000481_ (accessed 10 April 2017).
11 Foreign Policy Concept for 2014–2020, Ministry of Foreign Affairs, Republic of Kazakhstan. Available at: https://web.archive.org/web/20161027061432/http://mfa.gov.kz/

index.php/en/foreign-policy/foreign-policy-concept-for-2014-2020-republic-of-kazakhstan (accessed 26 July 2017).

12 Kazakhstan and Russia signed the 2010 Agreement on Common Use and Protection of Transboundary Rivers. A joint water committee and six working groups have been set up to regulate shared rivers, namely the Ural/Wula'er, the Irtysh, the Ishim/Yiximu and the Tobol/Tuobo'er Rivers.

13 Integration of Central Asia – way to solve regional water and energy problems, *News Central Asia*, 10 October 2016. Available at: www.newscentralasia.net/2016/10/10/integration-of-central-asia-way-to-solve-regional-water-and-energy-problems/ (accessed 10 April 2017).

14 Interview with Cheng Yang, Chinese expert on Sino-Kazakh relations, 3 May 2016.

15 SCO members include central Asian countries, namely China, Russia, Kazakhstan, Tajikistan, Kyrgyzstan, Turkmenistan and Uzbekistan, while India is an observer nation.

16 President Nursultan Nazarbayev proposed the creation of a joint Water Committee at the SCO Summit 2014.

17 See script for Foreign Ministry Spokesperson Liu Jianchao's Press Conference on 14 February 2006. Available at: http://dk.china-embassy.org/eng/fyrth/t235451.htm (accessed 30 August 2017).

18 News, China attends to Kazakh interests over cross-border river utilization, protection, 29 January 2010. Available at: http://en.people.cn/90001/90776/90883/6882187.html (accessed 30 August 2017).

19 China allocates Ili to Kazakhstan, *Xinhua Net*, 4 July 2014. Available at: http://hy.cebnet.com.cn/2014/0704/265226.shtml (accessed 30 August 2017).

20 See a declaration signed between China and Kazakhstan, entitled 'The cooperation between the Republic of Kazakhstan and the Peoples Republic of China in the field of use and protection of transboundary water resources' (2014).

21 More than 30 unique cold-water fish have been found to live in this international river.

22 See: http://msn.huanqiu.com/world/hot/2012-06/2825013.html, 15 June 2012 (accessed 10 April 2017).

## References


Adeel, Z. and Wirsing, R.G. 2016. *Imagining Industan: Overcoming Water Insecurity in the Indus Basin*, Switzerland: Springer.

Biba, S. 2014. Desecuritization in China's behavior towards its transboundary rivers: The Mekong River, the Brahmaputra River, and the Irtysh and Ili Rivers, *Journal of Contemporary China*, 23(85): 21–43.

Blua, A. 2004. Central Asia: China's mounting influence, part 3 – Xinjiang's thirst threatens Kazakh water resources, *Radio Free Europe/Radio Liberty*, 18 November. Available at: www.rferl.org/a/1055958.html (accessed 12 July 2017).

Chen, Y. ed. 2014. *Water Resources Research in Northwest China*, Dordrecht: Springer.

Clarke, M. 2014. Kazakh responses to the rise of China: Between elite bandwagoning and societal ambivalence, in N. Horesh and E. Kavalski, eds, *Asian Thought on China's Changing International Relations*, Basingstoke: Palgrave Macmillan, pp. 141–172.

Deng, M. 2010. Research and thinking on water resources in Xinjiang [Xinjiang shui ziyuan wenti yanjiu yu sikao], *Quaternary Sciences*, 30(1): 107–114.

European Dialogue. 2011. Conflict of Interests Has Created Water Crisis, *European Dialogue*, 13 April. Available at: www.eurodialogue.eu/conflict-interests-has-created-water-crisis (accessed 12 July 2017).

Granit, J., Jägerskog, A., Löfgren, R., Bullock, A., Gooijer, G. de, Pettigrew, S. and Lindström, A. 2010. *Regional Water Intelligence Report Central Asia, Baseline Report*, Paper 15, Stockholm: UNDP Water Governance Facility.

Haftendorn, H. 2000. Water and international conflict, *Third World Quarterly*, 21(1): 51–68.

Jia, S. and Liu, J. 2014. *China's Water: A Report on Water Issues* [*Zhongguo shui wen ti bao gao*], Wuhan: Huazhong University of Science and Technology Press.

Li, M. and Li, H. 2014. Solution to water demand in the urbanization and industrialization of Xinjiang [Xinjiang Xinxing Gongyehua, Chengzhenhua de Shuiziyuan Duice], *Macroeconomic Management*, 11: 80–82.

Liang, C. 2011. Joint water diversion project launched in Xinjiang, *China Daily*, 15 April. Available at: www.chinadaily.com.cn/china/2011-04/15/content_12336268.htm (accessed 21 June 2016).

Stone, R. 2012. Transboundary rivers for China and Kazakhstan, no meeting of the minds on water, *Science*, 337(6093): 405–407.

Tai, Q., Yang, J. and Zang, J. 2003. The environment and ecological issues in the exploitation of Xinjiang's oil and gas [Xinjiang Youqi Ziyuan Kaifa Shengtai Huanjing Wenti Yanjiu], *Tribune of Social Sciences in Xinjiang*, 1: 29–34.

Tang, Q. and Li, L. 1999. Water resources characteristics and sustainable development of main international rivers in Northwest China, *Acta Geographica Sinica*, 66(S1): 21–28.

Tengrinews. 2013a. Turning rivers back into Kazakhstan may solve problems, *Tengrinews*, 27 September. Available at: https://en.tengrinews.kz/environment/Turning-rivers-back-into-Kazakhstan-may-solve-problems-22921/ (accessed 22 June 2016).

Tengrinews. 2013b. Kazakhstan and China discuss construction of waterworks facilities at transborder rivers, *Tengrinews*, 18 January. Available at: https://en.tengrinews.kz/politics_sub/Kazakhstan-and-China-discuss-construction-of-waterworks-16104/ (accessed 22 June 2016).

Tir, J. and Ackerman, J.T. 2009. Politics of formalized river cooperation, *Journal of Peace Research*, 46(5): 623–640.

Tursun, N. 2013. Rivers threatened as China, Kazakhstan water pact remains elusive, *Radio Free Asia*, 24 June. Available at: www.rfa.org/english/news/uyghur/delay-06242013164251.html (accessed 17 June 2016).

UNDP (United Nations Development Programme). 2011. *Intergovernmental Water Relations in Ile-Balkhash Basin*. Available at: www.cawater-info.net/bk/water_law/pdf/review_ile-balkhash_eng.pdf (accessed 12 July 2017).

UNDP Kazakhstan. 2007. Integrated water resources management and water efficiency in the Republic of Kazakhstan for 2008–2025, Progress report, United Nations Development Programme Project No. 35289, Annex 1, Draft. Available at: www.gwp.org/globalassets/global/toolbox/about/iwrm/asia/kazakhstan-iwrm-action-plan.pdf (accessed 12 July 2017).

UNECE (United Nations Economic Commission for Europe). 2009. *River Basin Commissions and Other Institutions for Transboundary Water Cooperation: Capacity for Water Cooperation in Eastern Europe, Caucasus and Central Asia*, New York: United Nations Publications.

Vogtmann, H. and Dobretsov, N. 2003. *Transboundary Water Resources: Strategies for Regional Security and Ecological Stability*, Dordrecht: Springer.

Wang, Q., Zhou, X. and Hu, X. 2007. Consultation mechanism framework of inter-regional water resources management in China, *Advances in Science and Technology of Water Resources*, 27(5): 80–84.

Wang, W. and Jia, J. 2016. Silk road economic development: Vision and path, in A. Amighini, ed., *China Dream: Still Coming True*, Novi Ligure: Edizioni Epoké.

Xing, G. 1996. *China's Relations With the Newly Independent Central-Asian Countries* [*Zhongguo yu zhongya guojia de guanxi*], Haerbin: Heilongjiang Education Publisher.

Yoffe, S., Wolf, A.T. and Giordano, M. 2003. Conflict and cooperation over international freshwater resources: Indicators of basins at risk, *Journal of the American Water Resources Association*, 39(5): 1109–1126.

Zhang, N. 2015. Позиции Казахстана в трансграничном сотрудничестве по использованию водных ресурсов [Principles of Sino-Kazakh cooperation on transboundary water resource management/Hasakesitan kuajie shuiziyuan hezuo jiben lichang fenxi], *Russian Central Asian & East European Market*, 4: 17–28.

Zhou, H. and Zhang, J. 2005. Analysis on the volume of available water resources and its carrying capacity in Xinjiang, China, *Arid Land Geography*, 28(6): 756–763.

Zhou, H., Song, Y. and Chen, X. 2003. Water utilization and related ecological problems, in X. Pan, W. Gao, M.H. Glantz and Y. Honda, eds, *Ecosystems Dynamics, Ecosystem-Society Interactions, and Remote Sensing Applications for Semi-arid and Arid Land*, Bellingham, WA: SPIE.

# 8 North-East Asia

## Sino-Russian cooperation over the Amur River

### Introduction: Sino-Russian relations

China and Russia, two continental superpowers, have the historical burden of ideological and political disparities. Diplomatic relations can be traced back to the early seventeenth century, when conflict was shown between Russia and China after Russia took possession of eastern Siberia. Bilateral relations have improved greatly since the 1920s when the Soviet Union exerted more influence over the young Republic of China. For various reasons, among which political leadership under Mao is a strong factor, mutual ties developed rather slowly over the ensuing decades, even after 1949 when the People's Republic of China was established. Until the late 1950s, the divergence between the two countries' leaderships grew wider, ultimately creating a Sino-Soviet split, where bilateral ties and communications were abrogated for the next three decades. It was not until the 1990s when Sino-Russian relations resumed normality. With a proliferation of institutional diplomatic cooperation in the 1990s, the two countries announced the development of their 'strategic partnership' at the turn of the century.[1] Sino-Russian relations have displayed dynamic changes and have been regarded as ambiguous and uncertain, with some even viewing the bilateral relations and the rationale and motives behind them as being full of contradiction (Lo, 2008). Although since the early 2000s the two countries have not developed a defensive strategic partnership (Fu, 2016), a very sensitive military–technological cooperation between China and Russia has been formed, demonstrating a strategic aspect of Sino-Russian relations.

China and Russia share a transboundary river, the Amur River, which rises in Mongolia and then flows through the Russian Far East (RFE). Its tributary, the Ussuri River, marks the border between China and the Russian province of Primorsky Krai. The length of the Amur watercourse is 4,440 km, of which 2,800 km passes through RFE territory. Since the 1990s, China and Russia have formed progressive diplomatic cooperation in this region, and this is greater than the level of cooperation they have developed for their other rivers, such as the Ob in central Asia. Along with increased industrialization and urbanization, however, both countries have focused increasing attention on the use of the Amur River. They have each developed their own agenda for exploiting economic interests from the river. As such pursuits are driven by different interests, the sharing of the Amur River has become a source of disagreement between the two countries.

Through institutional cooperation, China has developed diplomatic initiatives with Russia on the sharing of the Amur River. With increasing population growth and industrialization, China has shown greater demands for the exploitation of the Amur River, including for agricultural irrigation, hydropower generation and navigation. China has also shown its strategic interest in the RFE by developing the 'One Belt, One Road' initiative that involves the exploitation of the Amur River and a broader interest to expand trade and investment in the RFE. The RFE is an important transit route between Western Europe and countries in the Asia-Pacific region due to its year-round access to the sea and direct connection with the Trans-Siberian and Baikal–Amur railways.[2] The RFE's advantageous geographical location has made it significant to China's social and economic development. As part of the Silk Road Economic Belt[3] initiative, Heilongjiang Province, where the Amur River runs through China, has its own 'Heilongjiang Land and Maritime Silk Road Economic Belt' initiative and is also keen on accessing the Asia-Pacific through the twenty-first-century Maritime Silk Road. Heilongjiang Province needs the RFE as a transport corridor to the Asia-Pacific, which it assumes requires the economic integration of China's North-East (Dongbei) and the RFE. However, mistrust remains between the two countries, and the Russians have suspicions and long-held fears that China intends to reclaim the former Chinese territories, given that the increased growth and trade opportunities in the Amur River area have attracted increasing numbers of Chinese migrants.

Furthermore, Russia has exhibited its own rationale in the management of water resources, and this somewhat differs from that of China. How Russia will respond to China's diplomatic initiatives remains unclear. It has, in the past, adopted unilateral plans to construct dams along the Amur River to better control the flow of water and reduce the impacts of flooding. As the Putin administration has placed high importance on the exploitation of the Amur River and the RFE, seeking hydropower has become incorporated into Sino-Russia energy cooperation.[4] Russia has become more determined to strengthen trade and policy relations with China, including through cooperative initiatives such as energy generation (Xinhua, 2014; Paik, 2015). This reflects the high importance the Putin government has placed the exploitation of the Amur River and the RFE. However, questions are raised as to how cooperation between the two countries will integrate the New Silk Road with Russia's Eurasian Economic Union project, a regional integration project that aims to unify the former Soviet republics into a single economic and political space.[5] Much of its economic integration is based on the exportation of natural resources (Sadri, 2014), and the wise use of shared water resources.

What makes this issue particularly sensitive is that the Amur River is an important symbol of, and geopolitical factor in, Chinese–Russian relations. Historically, China and Russia have experienced serious border disputes. Border clashes occurred between the two countries as a result of political divergence in the 1960s. In border agreements signed in the 1990s, the Amur River was adopted to demarcate the boundary between the two countries, leaving a number of Amur River islands out of discussions. Without determining who these islands belong to, the Sino-Russian border disputes remained unresolved (Zinberg, 1996).

Therefore, China's water diplomacy with Russia has been complicated by the differences between its own and Russia's strategic interests over the Amur River. Although Sino-Russian cooperation has shown a positive trend, questions have been raised about the future of the Russia–China partnership. China's water diplomacy has strong implications for the development of the RFE and future bilateral relations between the two countries. This chapter examines the interactions between China and Russia regarding the sharing of the Amur River. It investigates the complex nature of water-sharing, including issues surrounding border disputes, economic development and water management. The chapter concludes by assessing the outcomes of water diplomacy and predicting its future development.

## River characteristics

China and Russia have a shared border stretching for more than 4,300 km, and China's Heilongjiang Province and Russia share nearly 3,000 km of the border (see Figure 8.1). The Heilongjiang River or Amur River is a boundary river with

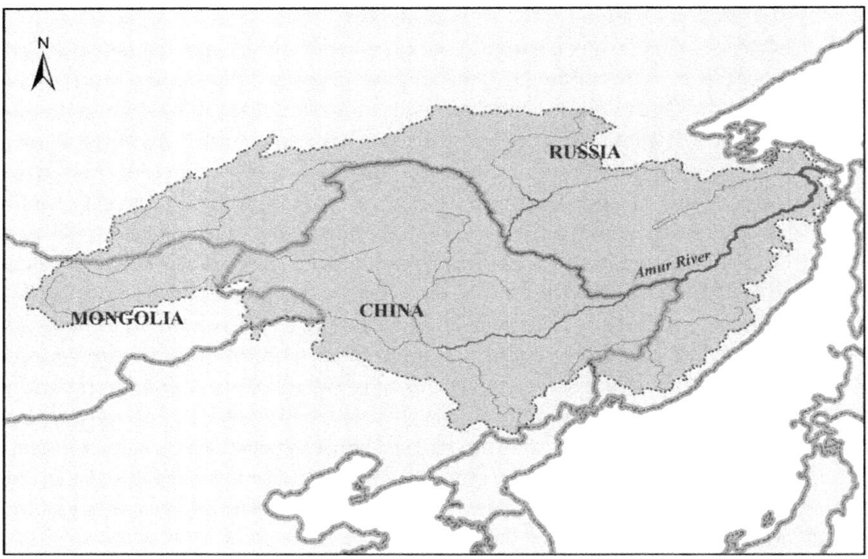


*Figure 8.1* The Amur River basin.

Source: authors' commissioned work, credit: Jiabao Yan.

complex features. Originating in the Khentii Mountains of Mongolia,[6] the Amur River is the world's tenth longest river with a total length of 4,370 km. Much of this length forms the border between China and Russia. The river flows eastwards until it connects with the Strait of Tartary on Russia's east coast. The Amur River basin is shared by four countries, being Russia, China, Mongolia and North Korea. Its total basin area is 2,085,841 km$^2$. The river features an uneven seasonal and annual flow, with summer floods that account for up to 80 per cent of its annual flow. The upstream and middle range of the Heilongjiang River are located at the border between China and Russia.

### Water diplomacy and border disputes

The shared Amur River runs through a disputed border area which has long been the subject of heated debates. Only in the early years of the twenty-first century has the border disagreement been settled. During the Qing Dynasty (1644–1912), the Amur region remained a relative backwater of the Qing Empire, with Aigun being a major town on the river. When the Qing Dynasty began to weaken, Russia forced the Manchus to yield all lands north of the river to the Russian Empire by the Treaty of Aigun (1858) and it also gained possession of the tributary territories of the Amur and Ussuri rivers. With the Convention of Peking (1860), the Russian Empire and the Qing Dynasty agreed that the border would consist of the ridge of the Sarikol Range, but the exact border remained contentious throughout the twentieth century.

In the 1960s, along with the decline in foreign relations between China and Russia, critical disagreements occurred over the ownership of the Ussuri River's midstream islands. China claimed these islands as they were located on its side of the river (if demarcated according to international shipping lane rules). The USSR wanted (and by then, already effectively controlled) almost every single island along the river. On 1, 14 and 15 March 1969, Soviet and Chinese forces clashed near Damanski/Chenpao Island on the Ussuri River. Since then the two countries have debated the demarcation of the borderline. One issue is that the ownership of many islands remains unclarified, and Heixiazi Island is one of these. In 2008, as a gesture for peaceful settlement, Russia transferred Yinlong (Tarabarov) Island, half of Heixiazi (Bolshoi Ussuriysky) Island and Bolshoi Island on the Argun River in the Chita Region of East Siberia to China as part of an addendum to an agreement on the Russian–Chinese border. The territorial area of Heixiazi Island that China now occupies is slightly less than that of the Russian side (see Figure 8.2 and Box 8.1). This agreement concluded longstanding negotiations that were launched by the Chinese government in 1964.

### River basin security

The Amur River is faced with pressing issues relating to the increasing exploitation of its water resources. The river's three riparian states – China, Russia and Mongolia – have shown varying degrees of reliance on the river, which relates to their

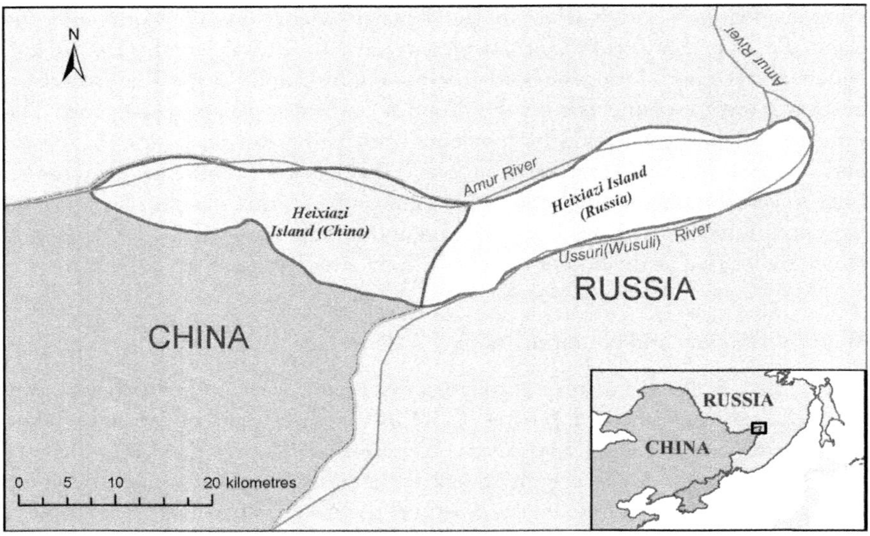


*Figure 8.2* The geographical area of Heixiazi Island.

Source: authors' commissioned work, credit: Jiabao Yan.

## Box 8.1 Heixiazi Island as a disputed area

Tarabarov/Yinlong Island and the Bolshoi Ussuriysky/Heixiazi Island are located at the confluence of the Heilong/Amur and Wusuli/Ussuri Rivers. The confluence of these river systems has created the 327 km$^2$ oasis. The island guards the rivers' navigable throat of the Amur-Ussuri River and is also close to the Russian city of Khabarovsk, which lies just off its north-eastern tip. The island is flat and open, and supports a good vegetation cover, with swamps, woods, shrubs and lush grasslands. It is home to a large number of species such as black storks. The geographical location is crucial for a number of Russian cities in the region of Khabarovsk Krai.

Since the 2008 settlement, the Chinese government advocates a progressive development of the island that combines the conservation of natural resources with opportunities for trade and economic development. It has won Russia's consent to construct transport infrastructure such as a port on the Amur waterway and a linked Heixiazi highway network that will include the Tongdao Bridge. Similar construction is initiated on the islands on Ussuri River as well. A recreational zone that promotes ecotourism has been established on the Heixiazi Island. In a later stage, an industrial park is also planned to facilitate enhanced trade between both countries. This park is planned to house factories for manufacturing products for export to the Russian market. However, as China is taking a lead to exploit resources on its side of the island, Russia fears that such works will affect water quality on its half of the island.

differing levels of population and economic growth, and results in differing levels of river planning and management among the three countries (see Table 8.1).

For China, agricultural consumption accounts for over 75 per cent of the total water usage in Heilongjiang Province (see Table 8.1) with most of the water being consumed for rice production. With over 93 per cent of the Amur River basin's population residing within China, China has demonstrated a significantly higher water resource utilization rate (water quantity adoption divided by resources that it possesses in its territory) than the other two riparian states, reaching an alarming level of 41 per cent (Yan, forthcoming).[7]

The remaining 7 per cent of the basin's populace is sparsely distributed between Russia and Mongolia. Most of the population reside in Russia, and so it thereby consumes more water from the Amur River than Mongolia. Russia's total withdrawal now amounts to approximately 5 per cent of the total withdrawal of the basin. Agricultural irrigation accounts for over two-thirds of its total withdrawal. In comparison, only a small proportion of the Mongolian population lives by the river, and the country's withdrawals from the basin are therefore at a negligible 1 per cent.

With respect to economic growth, both China and Russia are relatively better off than Mongolia in GDP per capita (see Table 8.1). However, the per capita GDP reached in places along the river does not reach the average GDP in neither China nor Russia. With the Amur River being an important river to national policy, that the output in water management along the river basins is not much different from that at national level respectively. This indicates that the rate of economic growth is not a significant factor impacting on the peaceful resolution of issues in the sharing of water resources between China and Russia.

### *Concerns in the sharing of the Amur River*

As discussed above, China and Russia strongly contrast in their reliance on the Amur River. China is experiencing rapid population growth and industrialization in Heilongjiang Province. With its rapid urbanization in Heilongjiang Province, the country utilizes a significantly greater share of the river than Russia and a number of disastrous events have occurred that have had impacts on the downstream of the Amur River in Russia. These events include the 2008 oil spill on the Amur, the 2010 Chemical Barrel Incident on the Songhua, and the 2013 flooding event that affected both countries. Although Russia has not openly displayed disagreements with China, the country has shown concerns about how China's planning and management impacts on the shared river. Such concerns relate to pollution control, disaster prevention and the protection of biodiversity.

Water quality in the Amur River has been a growing concern, as risks of river pollution have also grown. The river basin has experienced a range of issues that threaten water quality, including chemical pollution, bank erosion and reductions in floodplain land. It is feared that pollution from the Songhua River could affect the Amur's water quality in Russian territory. However, some consider that Russia is not significantly exposed to the threat, given its distance from the

*Table 8.1* China and Russia's water resource management in the Amur River basin

| | *River catchment area ($km^2$)* | *Runoff depth (mm)* | *Water resources ($10^9 m^3$)* | *Agricultural water withdrawal ($10^9 m^3$)* | *Industrial water withdrawal ($10^9 m^3$)* | *Domestic water withdrawal ($10^9 m^3$)* | *Total withdrawal ($10^9 m^3$)* | *Water resource utilization rate (%)* | *GDP per capita (US$)* |
|---|---|---|---|---|---|---|---|---|---|
| China | 887,411.6 | 126.47 | 1,122.3 | 321.5624 | 91.22827 | 41.45740 | 454.2481 | 40.5 | 447 |
| Russia | 1,009,186.36 | 226.80 | 2,288.8334 | 3.6459 | 3.3669 | 12.4742 | 19.4870 | 0.9 | 2,701 |
| Mongolia | 289,243.82 | 28.46 | 53.8666 | 0.35233 | 0.1037 | 0.1426 | 0.5986 | 1.1 | 1,836 |

Source: Yan (forthcoming).

pollution source and the dilution of any pollutants that will occur before they reach Russian territory (Watts, 2005).

The risk of flooding on the Amur River has been a concern since formal Sino-Russian foreign relations were resumed. As early as 1986, an informal agreement was reached for local Russian authorities of the Amur region to work with China to better control floods on the watercourse. There have been a number of emergency incidents over recent years, such as the 2013 flooding[8] that impacted 52.47 million people in Heilongjiang (Simonov *et al.*, 2016). Other flooding events have occurred on the upstream and midstream parts of the river and there have also been several ice flood events (Luo, 1996; Wang *et al.*, 2015). Such incidents have further strengthened the two countries' willingness to commit resources for the prevention of natural disasters.

The Amur River is rich in biodiversity including abundant fisheries resources (Zou *et al.*, 2014). The river spans five distinct ecological regions, from grasslands and boreal forests to desert and tundra. It is famous for its diversity of flora and fauna genus and species, and the ecosystems of its river valleys and mountain divides. More specifically, the basin contains 17 wetlands, all with immense flora, fauna and aquatic life. Within the Russian expanse of the Amur River alone, there are at least 2,800 and 500 plant and animal species respectively, including critically endangered animals such as the Siberian Tiger (Voronov, 2004). In the Amur River basin, relic species are also found in the river, such as vascular plants, birds and mammals.[9]

The Amur River basin also hosts a rich diversity of wetlands. Of the 220 nature reserves in the China segment of the Amur (Heilong) basin, 83 are wetland reserves (37 per cent of natural reserves for the region and nearly 24 per cent of all wetland reserves in China). Six of the national nature reserves in the Amur (Heilong) basin are also listed as Ramsar sites (i.e. those at Xingkai Lake, Sanjiang, Honghe, Zhalong, Dalai Lakes and Xianghai). Many of the remaining wetlands in the region are designated as nature reserves at the national, provincial or county level. Concerns exist that there is a lack of transboundary coordination to protect the wetlands in the river basin.[10] It is feared that industrial discharge and spills, and associated degraded water quality, are direct threats to the protection of the river's flora and fauna (Buznik, 2001).

## Sino-Russian water diplomacy

### *Principles in water diplomacy*

In this section, the positions and principles of Sino-Russian water diplomacy are examined before we give an analysis of the content and features of institutional cooperation. To facilitate its existing water planning and management policies, China has developed bilateral cooperative initiatives with Russia on the sharing of the Amur River.

Foreign relations between China and Russia have laid the foundations for both countries to assume diplomatic cooperation on water sharing (Vinogradov and Wouters, 2013), even if this cooperation has been sporadic. In the Treaty of Good-Neighbourliness and Friendly Cooperation (2001) between China and Russia, Article 19 states that both sides are willing to cooperate on the protection and improvement of the environment. More specifically, this chapter states the cooperation is for the:

> prevention of cross-border pollution, the fair and rational use of water resources along the border areas, and the use of biological resources in the Northern Pacific and boundary river areas; to make joint efforts in protecting rare flora, fauna and the natural ecosystem, and conduct cooperation in preventing the outbreak of major accidents arising from natural disasters or technical reasons and eliminating their after-effects.

This may explain why contained responses were made from both countries when disastrous events occurred on their shared rivers. Since the mid-1980s, water diplomacy has also resulted in various forms of institutional cooperation, including memoranda of understanding, treaties and protocols (see Table 8.2).

Russia is a participant in the UNECE,[11] meaning that it agrees to adopt the norm of developing collaborative water management with neighbouring countries that it shares water resources with. However, the country has developed no systematic basin-wide water management planning for its shared rivers. Instead, Russia has largely followed China's approach in developing bilateral cooperation, and only in rare cases have multilateral agreements been developed that incorporate Mongolia. Only in 1994 have Russia, China and Mongolia signed an agreement to establish the Dauria International Protected Area. However, more collective efforts have been formed between China and Russia.

China has complied with certain international norms although it is not a participant in the UNECE. China recognizes that transboundary impact can be caused by disastrous and emergency events, and has therefore adopted the principle of 'no significant harm' which was established in the 1997 United Nations Watercourse Convention. Both China and Russia have agreed that they would commit 'to develop and adopt necessary measures aimed at preventing and reducing transboundary impact on transboundary waters' (Vinogradov and Wouters, 2013). The agreement facilitates similar coordination of both countries when faced with a significant transboundary impact.

Data sharing is another emphasized aspect of the institutional cooperation between China and Russia, with both countries agreeing that information-sharing should be conducted in a transparent and timely fashion. According to the 2008 agreement between China and Russia on Reasonable Utilization and Protection of Transboundary Waters, Article 5 states that 'Both sides agreed on the content, quantity and timing of the exchange of information on transboundary water through consultation. In the event that one party proposes to exchange non-agreed information or data, the other shall comply when circumstances permit'.[12] Such an agreement has generated greater confidence between the two countries and eliminated the suspicion of water information being used

*Table 8.2* Institutional cooperation between China and Russia

| *Date* | *Name of agreements* | *Sectoral interests* |
|---|---|---|
| 1986 | Agreement on development of the Amur transboundary water management scheme | The protection of water ecology; and the exploitation of water resources |
| 1988 | Agreement on the protection, regulation and breeding of bio-resources in the transboundary Amur (Heilong) and Ussuri (Wusuli) Rivers | The protection of water ecology and fish stocks |
| 1994 | Agreement on the protection, regulation and breeding of bio-resources in the transboundary Amur (Heilong) and Ussuri (Wusuli) Rivers | The protection of water ecology and fish stocks |
| 1994 | Russia–China Cooperation in Protection of the Natural Environment | The protection of the environment in general |
| 1994 | Agreement on the Dauria International Protected Area | Biodiversity protection |
| 1996 | Agreement on the Khanka/Xingkai Lake International Nature Reserve | The protection of water ecology and nature reserves |
| 2001 | Russia–China treaty on good neighbour relations, friendship and cooperation | The protection of water ecology; and the exploitation of water resources |
| 2002 | Agreement on cooperative monitoring of water pollution in the Amur-Heilong River | Water pollution control |
| 2006 | Protocol on 'common approaches to monitoring of transboundary waters', namely the Amur (Heilong), Ussuri (Wusuli), Argun and Razdolnaya (Suifen) Rivers | water quality control |
| 2006 | Agreement between China and Russia regarding cooperation on prevention and elimination of emergency situations | Emergency and disastrous event control |
| 2007 | Integrated management of the Amur basin, between Khabarovsk Krai and Heilongjiang Province for the purpose of protection of transboundary water ecosystems, within the Programme of Cooperation Between the Regions of the Far East and Eastern Siberia and the North-East of the People's Republic of China (2009–2018) | Protection of transboundary water ecosystems |
| 2008 | Agreement between China and Russia regarding cooperation on rational management and protection of transboundary waters | The protection of water ecology |
| 2011 | Sino-Russian Joint Commission on Border River Shipping | Navigation and transport |
| 2011 | Sino-Russian Strategy for the Development of a Transboundary Network of Protected Areas in the Amur River Basin for the Period till 2020 | Inventory and protection of wetlands |
| 2013 | Memorandum of intent to strengthen Sino-Russian cooperation around natural disasters, particularly with regards to aid and information provision | Emergency and disastrous event control |
| 2016 | Memorandum on transboundary cooperation between 16 Chinese and 12 Russian protected areas located along the Amur and Ussuri Rivers on the Sino-Russian border | The protection of water ecology and biodiversity |

as 'weapon'. Both countries share information on the state of the Amur's river beds as well as regarding risk of flood. Based on this strong confidence, both countries have responded to the other's information requests and both countries have benefited from the collaboration by reducing losses associated with flooding, such as reduced loss of human life and infrastructure damage.

### Areas of collaboration

Through China's water diplomacy, a piecemeal approach has been adopted in its water management activities. Most of the efforts reflect an end-of-pipe approach to issues that arise from water-sharing, including quality control, the protection of biodiversity along rivers basins, and disaster management.

#### *Quality control*

Since 2000, China's water diplomacy has developed a rather narrow focus over the control of water quality in its shared rivers. Early collaboration was conducted on the Amur (Heilong) and Ussuri (Wusuli) Rivers, where three locations were selected for monitoring.[13] The monitoring included samples of water and benthic animals collected from the left and right banks, and monitoring of hydrological parameters such as flow rate, depth of water and river width. Joint measurements were renewed following the 2005 incident,[14] where an enlarged scope of investigation was agreed to cover five rivers and nine locations.[15]

However, negotiations have not been smooth between China and Russia. It has been noted that both China and Russia lack the coordination to further cooperate to regulate water quality, as there is a paucity of collectively adopted standards and regulatory tools in water management.[16] The Russian side has raised issues about how to resolve accidents that result from China's weakened management. The Russian representatives have also expressed their interest in China providing compensation for the pollution that has emanated from within Chinese territory. As a result, at national level, diplomatic cooperation has been slow. Disagreements have occurred over what regulatory standards are to be adopted, as well as the allocation of work in the management of water resources.[17]

In policy implementation, it is not surprising that China had to take unilateral action to improve the water quality of the Songhua River (Shi, 2006; see also ADB, 2016a, 2016b). The central government has increased funding to reduce the future risks of pollution in this river as it flows through both Jilin and Heilongjiang Provinces. This funding was for projects on, for instance, sewage treatment (BBC, 2006; Shi, 2006; Li and Tian, 2010). By so doing, the Chinese government aims to reduce the risks to Russian water quality and the associated potential for strained relations.

It is also noteworthy that there is a breakdown between national and provincial efforts. In China's Eleventh Five-Year Plan, Heilongjiang Province has taken the initiative to improve the ecological protection of the shared rivers.

For instance, the province has increased its efforts to improve the function of water conservation areas, to strengthen the protection of urban drinking water sources, reduce soil erosion and non-point source pollution, and to prevent pollution and improve water quality in the border river basin by building sewage treatment plants. On the Russian side, promoted by local authorities, Khabarovsk Krai and Heilongjiang Province have formed an agreement to conduct integrated management of the Amur basin. This agreement involves the joint monitoring of surface water quality and aquatic habitat, and the creation and effective management of joint conservation areas for the purpose of protecting the transboundary water ecosystems.

### *Water ecology and biodiversity protection*

China's high level of water use has raised concerns that insufficient water may be available for various downstream functions of the river, such as ecological protection. Cooperation to address concerns about ecological degradation and the loss of biodiversity throughout the Amur River basin has been underway since the 1980s. Along with a shift to increased water-quality monitoring, growing emphasis has been placed on the protection of the aquatic ecosystem and China and Russia have now adopted a series of cooperative initiatives in these areas. For instance, they have signed the 'Sino-Russian Strategy for the Development of a Transboundary Network of Protected Areas in the Amur River Basin for the Period till 2020', where inventory and protection of wetlands are placed as priority (2011). In 2010, Chinese and Russian officials agreed to set up their very first protected area for an endangered species, the Amur tigers, also known as Siberian tigers, which have their habitat between Jilin Province in China and neighbouring Primorsky Province in Russia. This new transboundary protected area was to provide a wider and healthier habitat for Amur tigers and other endangered species, such as the Amur leopard, musk deer and goral. Such efforts have also reflected China's domestic policy emphasis on wetland conservation. There is also an increasingly institutionalized preference for restoration projects over the protection of pristine natural habitats. This arises in part because of government subsidies for returning farmland to wetland. This is reflected in the 2016 Memorandum on Transboundary Cooperation between 16 Chinese and 12 Russian protected areas located along the Amur and Ussuri Rivers on the Sino-Russian border. The total area of these partner reserves comprises 1.9 million ha.

It is noted that the impetus for Sino-Russian cooperation in managing the Amur River basin comes primarily from the Chinese side (Vinogradov and Wouters, 2013). For instance, while China has implemented a US$1.9 billion programme to provide environmental remediation of the Sungari River, which is the largest tributary of the Heilongjiang River (Simonov and Dahmer, 2008), no comparable measures have been taken on Russian side (Vinogradov and Wouters, 2013). Additionally, China was the driving force in establishing a joint committee to promote annual communication between the two countries.

Since then, strategic plans have also been developed and implemented, with transboundary nature reserves being built. Among a series of six of the type, Lake Khanka-Xingkai International Nature Reserve is one of the earliest established. Both countries have now also agreed to conduct joint visits, exchange data and conduct collective educational programmes.[18]

The state-led Sino-Russian transboundary water cooperation has now recognized the role that the epistemic community and NGOs have played. Reduction of biodiversity in the RFE has been a key concern for a growing network of environmentalists and international actors (Buznik, 2001). Political opportunities have been provided to the environmental movement, particularly in Russia where the political system is relatively more open. Since around 2006, the Russian government has strengthened the administration of NGOs.[19] Domestic environmentalism within Russia is likely to articulate opposition, based on the movement's diversified interests in water resources. Such forces have had some success, such as in halting dam constructions over large rivers in Russia (Simonov and Dahmer, 2008). In particular, a network has been developed between a diversity of actors in Russia and China, incorporating the epistemic community as well as government authorities. Scientists and international NGOs have identified the pressing issue of the loss of wetlands in the Amur River basin, including within nature reserves, national parks, landscape and wildlife refuges and ecological corridors.[20] For instance, in implementing integrated management of the Amur basin, local government – the Heilongjiang government – has shown strong commitment. This has caught the attention of one of the local governments in Heilongjiang Province, which has since shown a strong commitment to implement integrated management of the Amur basin. It has shown much initiative in collaborating with Khabarovsk Krai and Heilongjiang Province to implement joint monitoring of the quality of surface waters and aquatic habitat.

### *Disaster prevention*

China and Russia have reached a few agreements (2006 and 2014) to collaborate on preventing water-related emergencies and flooding disasters. China and Russia agree that shared water information is important for their cooperation to be substantiated. The two sides have an equal agreement in data provision. Each provides water information from 14 water stations, covering water level, rainfall, and for some the level of water flow is also included (Wang *et al.*, 2015). Along with China's emphasis on disaster control, since 2013, the scope of data exchange has been expanded to include information to monitor potential floods and provide early warnings along the Amur River (Li and Tian, 2010; Xinhua, 2013a, 2013b). Indeed, improved cooperation was seen between China and Russia during the significant floods along the Amur River. Such cooperation makes it possible to undertake forecasting, regulate land uses in floodplains, protect floodplain wetlands and coordinate the construction of flood protection structures (Simonov and Dahmer, 2008).

It can be noted that a broader scope of information is increasingly shared between the two countries (Xinhua, 2013a, 2013b).

Water-related emergency management and associated human security is also strongly emphasized in China's water diplomacy. Collaboration has been extended to government departments such as between China's Ministry of Water Resources and the Russian Emergencies Ministry. Both countries have signed a peacemaking agreement which guarantees that refugees are provided necessary aid and that peacemaking troops and transit procedures are approved for managing disaster events. Furthermore, flexibility is allowed, as both sides agree to exchange hydrological data on the boundary water flow, water level and quality, ice conditions and boundary water protection measures as required to prevent flooding or ice hazards (Chu *et al.*, 2014; Wang, 2015). In implementation, frequent communication has been facilitated with a 'hotline' set up between the two countries' environmental ministries. Overall, China has been active in trying to minimize any potential negative impacts on Russia (Xinhua, 2005a, 2005b).

The two countries have also agreed that their collaboration for scientific research and capacity building should be strengthened to enhance responses to emergency events. In terms of institutional cooperation, measures have been developed in these areas, reinforcing the collaboration to disclose information and exchange necessary hydrological data. However, there is currently no integrated programme that incorporates planning for flood risk management and promoting the productivity and diversity of the common river ecosystem. It has been reported that Chinese flood-management efforts have only involved enhanced dykes, which shows the country has little inclination to consider ecological protection (Simonov *et al.*, 2016). From this perspective, the collaborative measures could be considered end-of-pipe type responses only.

With regard to implementation, unlike China's bilateral cooperation with other countries, no joint Sino-Russian committee has been established to coordinate management actions (Vinogradov and Wouters, 2013). Most of the collaboration has been developed by the national governments of both countries to establish specific working groups, such as in emergency management. Ministerial work is not coordinated and holistic water planning and management at the national levels has therefore not been facilitated.

### *River navigation and transport development*

Another focus of water management by both China and Russia is to profit from water resources by developing navigation channels in the two countries' shared rivers. An agreement to develop the watercourse was made in the late 1980s and early 1990s, facilitating the construction of a navigation route commencing at the Amur River and running for 894 km to its tributary. This navigation route connects the major Chinese cities of Ha'erbin, Jiamusi and Tongjiang which have already established trade relations with Russia. Further along the Ussuri River, the main port towns that are enjoying the established trade activities include Xiao'muhe, Hutou, Dong'an and Haiqing. These towns are also connected with the Argun and Shilka Rivers that join at Heihe city. Since 2004, intra-regional collaboration between Heilongjiang Province in China's north-east and the RFE has been revisited four times in Sino-Russian diplomatic cooperation (Shen, 2011). This collaboration indicates an expanded strategic partnership in Sino-Russian relations (Guo, 2013), particularly in infrastructure development (Chen, 2015). As the Russian central government is keen to improve infrastructure development in the RFE, it has become receptive to China's extended projects. Both countries have agreed on the development of improved inland water transport (Saritas *et al.*, 2015) to expedite related navigation and infrastructure projects.

China has been keen to promote infrastructure development on the rivers it shares with Russia. According to the Thirteenth Five-Year Plan of the Heilongjiang Province, cross-border bridges and a road construction project are essential to promoting regional development and realizing national government's 'One Belt, One Road' Strategy.[21] The new cross-border road bridge and its corresponding infrastructure is viewed as a foundation for opening up to Russia.[22] It has so far completed the construction of the Heihe and Dongning cross-border highway bridges which have served to improve Chinese investors' access to the RFE. Apart from the Europe Railway that links the city of Suifenhe in China's Heilongjiang Province with Russian cities, the navigation channel on the Suifen River is another important link with ports in the RFE.[23] These transit links have facilitated Heilongjiang's development into a regional connection hub, which is essential to promote its broader Eastern Silk Road Economic Belt strategy that is based on the province's crucial geographical location.

### *Dam construction*

Dam construction is another focus of Chinese water management policy. Since the country's first dam construction in the 1930s, many large reservoirs have been built for the primary purposes of hydropower generation and flood control. In Heilongjiang, there are 553 small reservoirs and dams, most of which have been developed on the Second Songhua River and its tributaries (Simonov and Dahmer, 2008). Among these, the Xiaofengman Dam is the largest.

China has strong incentives to undertake dam construction projects in its north-east region. Dam construction secures a water supply for urban areas in Heilongjiang Province and helps with meeting the region's growing domestic demand for electricity while also creating additional capacity for power exports from Siberia to China. In 2014, China imported 6.75 billion kWh of electricity (CIA, 2016), and approximately 1 per cent of this total was supplied by the RFE Federal District (FCS, 2014). Hence, large-scale hydropower development in the Amur basin has been planned for years among the river's riparian states including China, Mongolia and Russia. China's Twelfth Five-Year Plan aims to boost bilateral cooperation between China and Russia and to facilitate joint innovative development of both countries. Under this plan, a joint venture deal aims to build up to 10 GW of new hydropower capacity on the tributaries of the Amur and Songhua Rivers.

Russia's commitment to dam construction is much less than that of China. Historically, hydropower has been the main component of water infrastructure development in China, and the country ranks among the highest in the world for hydropower generation potential (IEA, 2010). In recent years, dam construction has become an important part of China's capital expansion (Simonov and Dahmer, 2008) and the industry has been dominated by a few private investors. At the same time, China's dam construction activities should be of interest to the Russian government for boosting development in the RFE, which is a region in demand of infrastructure investment as well as manpower (Zhou *et al.*, 2016).

Various negative effects from dam construction projects have been recognized. Environmentalists fear there will be further serious environmental impacts from planned new hydropower projects, many of which are certain to move forward given the enormous potential for power production in the Amur basin. The unnatural hydrological changes associated with dam constructions have caused a deterioration of the basin's floodplain wetlands. For instance, dams built near the Nen River, also called the Ni'erji Water Control Project, have already resulted in habitat loss and degradation, including the desertification of downstream wetlands, disruptions to seasonal fish movements and the loss of recharge capacity for the wetland aquifer (Simonov and Dahmer, 2008). It is also argued that the construction of the Zeya (built in 1975) and Bureya (built in 2003) hydroelectric reservoirs have already resulted in the destruction of large areas of forest and marshland. This has affected the ecosystems of the Amur River from a distance of more than 1,000 kilometres, disturbing the aquatic fauna and the mass-migration of Siberian roe deer, and flooding some regional and national protected areas (Parfitt, 2005; Simonov and Dahmer, 2008; Simonov *et al.*, 2016). Additionally, Indigenous populations report that the dams are responsible for changes in the local climate, water pollution as a result of rotting vegetation, and artificial periodic flooding of downstream settlements. Fishery depletion and the disruption of traditional navigation networks have created additional hardships for the people of the Amur who have long depended on the river for their way of life. The epistemic community has repeatedly warned about the ecological impacts of such dam construction projects. However, as such projects have been rapidly launched, it is unlikely that either China or Russia would benefit from incorporating the epistemic or local Indigenous communities into project planning.

There is uncertainty in Russian politics surrounding the issue of dam construction, particularly given that government decisions are significantly impacted by domestic opposition. Various discourses have been developed to legitimize dam construction policy and planning. In such discourse, as with China, dam construction is linked to facilitating improved flood control. In Russia, dams are advocated as helping to better control the flow of water and reduce the impact felt by any instances of flooding (TASS, 2014). For instance, the state-owned company RusHydro has proposed a 'flood control hydro power' plan involving four new dams on tributaries of the Amur River. Therefore, both China and Russia recognize that difficulties remain regarding how they can promote the development of large-scale hydropower projects on the main stream of the Amur River (Zhou *et al.*, 2016).

### Assessment of China's water diplomacy

Since the late 1990s and the resumption of Sino-Russian bilateral relations, China and Russia have engaged in increased collaboration for water management. Although in recent times Sino-Russian foreign relations have been relatively stable, China's development of water diplomacy with Russia has evolved relatively slowly. The issue of water diplomacy has a complex nature and it has been further complicated by ongoing disputes over border demarcations.

Sino-Russian institutional cooperation has been developed in various forms. Information sharing and 'no-harm' principles are adopted in diplomatic initiatives which provide a foundation for resolving potential disputes over shared rivers. Due to China's growing population and levels of urbanization, authorities are feeling increasingly threatened by the potential disasters the large international rivers may bring. As shown in other cases, China has also been proactive in negotiating with Russia regarding the collective exploitation of economic interests in transport, dam construction and other water-related trade activities. With the protection of human security being a strong focus in China's water diplomacy, the authorities were alerted by the large scale of damage caused by recent incidents.

Regular consultation and exchange of water data and sometimes flexible mechanisms have also been used. The development of collaborative water management is recognized by provincial governments in both China and Russia, showing efficiency in policy implementation. Provincial authorities in both countries have been incorporated to implement the agreements. Heilongjiang Province has taken a proactive position in liaising with its Russian counterparts to enhance cooperation. Additionally, enhanced cooperation has also been developed when a certain degree of decentralization is allowed by involving non-state actors. Biodiversity protection represents one of the few areas where collaborative efforts have been made, and where bilateral cooperation has been well implemented. Sino-Russian cooperation for biodiversity conservation has a long history and the Chinese government has placed an increasing focus on this issue in recent years. The incorporation of non-state actors such as environmental NGOs has also been helpful to speed up collaborative efforts to protect animal species in the Amur River basin. Therefore, it may be foreseen that in the long term, better collaborative water management can be made in other areas such as disaster control and the maintenance of water quality.

However, although the two countries have repeatedly confirmed their intention to work peacefully on water-sharing, on a few occasions, collaborative actions have not been well implemented. This is not too different from some of the other Sino-Russian cooperation projects, when agreements have also not been implemented (Dueben, 2013).

## Concluding remarks

China's water diplomacy has also encountered difficulties as both China and Russia have developed their own agendas in the sharing of the Amur River. With the bilateral relations being relatively stable, China has managed to develop a level of trust with Russia on its management of the Amur River. This has proved to reduce tensions when a number of disastrous events occurred that emanated from China's side. With agreement on sharing information to protect human security on both sides, there was barely any suspicion by Russia that the water data would be used by China for defence purposes. The Russian side is also convinced that China has been committed to reducing negative impacts on its domestic human security. From this perspective, we can say that China's water diplomacy with Russia has had some successes.

It is apparent that sharing water has become a politicized issue in China's foreign relations with Russia. China's commitment to reducing threats from water-related disasters is based on its interest to fulfil its regional economic development and foreign policy agendas. China is committed to proactively seeking opportunities to develop navigation, transport and hydroelectricity on the rivers it shares with Russia. With its water diplomacy, China has also initiated a link between water management and border demarcation issues. Its commitment to water management on the Amur River has the side effects of reassuring the Russian government of its commitment to building trustworthy neighbouring relations.[24] As a result, China has succeeded in negotiating its water management plans which has helped it to settle the border disputes with favourable outcomes.

China's strategic interests have become a concern for Russia, thereby constraining the country's commitment to collaborative water management. Mistrust remains on the border issue, with Russia expressing fears about China's increased investment in the RFE and that strategically, the two countries have developed rather different agendas for East Asia and Asia in general. For Russia, uncertainty remains regarding the extent to which it would commit to developing collaborative water management with China. Although regional development of the RFE has also gained attention in Russia's domestic and foreign policy agenda, due to the region's geographical location, the country has shown a weak capacity to implement its development plans. While the Russian national government has become more interested in collaborative exploitation of the Amur River as its strategic interests in the RFE have been substantiated, its agenda for regional development of the RFE differs from that of China. Therefore, Sino-Russian water diplomacy has shown its own characteristics, as the two countries have primarily focused on certain areas of common concerns, but such efforts remain sporadic and rather superficial.

It can be inferred that China's vested interests are not congruent with Russia's intention to tackle issues in the sustainable development of the Amur River basin. China's strategic interests in the RFE have illustrated a short-term nature that falls short of matching Russian interests. Therefore, China's rationale in prioritizing strategic interests on the issue of water sharing has shown limitations, which could hardly facilitate substantive bilateral cooperation between the two countries. The Chinese authorities may have been too optimistic about their water diplomacy with Russia, given that there is some reluctance emerging from the Russian side (Wu, 2015). Again, we are reminded that mistrust is a latent factor that cannot be ignored in Sino-Russian relations. China's prioritized strategic interest in the RFE has the tendency to impede its water diplomacy. To the Chinese authorities, the cooperative initiatives may remain unchanged. However, the outcomes would differ greatly if water management remains a secondary concern in Sino-Russian water diplomacy.

## Notes

1 In 2001, both countries signed the Treaty of Good-Neighbourliness and Friendly Cooperation Between the People's Republic of China and the Russian Federation.
2 These two routes are economically important for international passenger traffic and for transport of raw materials for international trade.
3 China has launched a 'Silk Road Economic Belt' development initiative. The aim is to promote integration between the countries of Eurasia. This is a strategic economic plan that aims to shift some of the domestic industrial overcapacity to neighbouring countries, particularly the resource-rich ones.
4 China and Russia had decided to develop energy cooperation since the late 1990s. Not until 2008 did both countries sign the framework agreement for a Memorandum of Understanding on Oil Cooperation, signalling accelerated cooperation in energy cooperation.
5 This union includes five states: Armenia, Belarus, Kazakhstan, Kyrgyzstan and Russia.
6 Although Mongolia is also located on Amur River, its portion of the shared river basin is rather small. The population that resides within Mongolia's share of the Amur River basin is also small, amounting to only 4 per cent of that of Russia's and 0.2 per cent of that of China's. The military and economic power of Mongolia is also much weaker than that of both Russia and China. Hence, Mongolia plays an insignificant role in affecting the management of the Amur River basin and its diplomacy with China and Russia will therefore not be considered in this chapter.
7 Generally speaking, 40 per cent is the threshold level for the sustainable management of the river basin.
8 Li Keqiang says China is to develop effective cooperation with Russia on flood control, *CCTV.com*, 2 August 2013, http://news.cntv.cn/2013/08/22/ARTI1377179693998402.shtml (accessed 9 February 2017).
9 Relic species of mammals include the scaly-sided merganser and Asiatic black bear. For more, see the Amur Information Centre Portal. Available at: http://amur-heilong.net/aic/en/1/04_econet_pas/0421HingganGorge/ (accessed 9 February 2017).
10 See report drafted by WWF Russia, 'Amur River concerns came to light at the Ramsar Conference'. Available at: www.wwf.ru/resources/news/article/eng/4320 (accessed 9 February 2017).
11 The Convention on the Protection and Use of Transboundary Watercourses and International Lakes (1992) is one of the most influential laws on the governance of

transboundary water resources. It advocates the formation of international cooperation among riparian states, for instance, the obligation to not cause harm.

12 China, Russia sign agreement on cross-border water use, protection, *Xinhua News*, 30 January 2008. Available at: www.china.org.cn/english/international/241291.htm (accessed 30 August 2017).

13 The three locations are Jiayin, Fuyuan and Tongjiang. See: China and Russia conduct collective activity to monitor water quantity and quality, *Xinhua News*, 8 November 2003. Available at: www.southcn.com/news/china/zgkx/200311080311.htm (accessed 9 February 2017).

14 A chemical plant explosion occurred in Jilin where the Songhua River runs through. Toxic pollutants drifted down the Amur River on the border between China and Russia.

15 The rivers are Xingkai Lake and the Amur, Argun, Ussuri and Suifen River.

16 China and Russia cooperate to protect the water resources in the Amur River, *Xinhua News*, 24 August 2007. Available at: http://gb.cri.cn/12764/2007/08/24/2945@1731873.htm (accessed 9 February 2017).

17 Ibid.

18 Ecological protection is well-developed on the Sino-Russian border, *Xinhua News*, 21 March, 2016. Available at: http://news.xinhuanet.com/world/2016-03/21/c_128816757.htm (accessed 1 July 2016).

19 See a list of Russian laws which indicate changes on the regulation of NGOs. Factsheet compiled by Katherin Machalek. Available at: https://freedomhouse.org/sites/default/files/Fact%20Sheet_0.pdf (accessed 30 August 2017).

20 For instance, the WWF Russia has played a leading role in the protection of wetland in the Amur River basin. See a letter addressed to the Deputy Secretary General of the Ramsar Convention on the issue. Available at: www.wwf.ru/resources/news/article/eng/4320 (accessed 1 July 2016).

21 The Thirteenth Five-Year Plan of the Heilongjiang Province. Available at: http://ghs.ndrc.gov.cn/ghwb/dfztgh/201607/P020160713591341270998.pdf (accessed 1 July 2016).

22 The road bridge infrastructure will be 19.9 km long. Some 6.5 km of the bridge and road junctions will lie in China, and the remaining 13.5 km in Russia. The length of the main suspension bridge will be roughly 1,300 m and its width 14.5 m.

23 Russia, China seeks to tap cooperative potential, *Xinhua News*, 6 September 2010. Available at: http://news.xinhuanet.com/english2010/china/2010-09/06/c_13481397.htm (accessed 1 July 2016).

24 Interview with Prof Cheng Yang, expert on Sino-Russian relations, 1 October 2016.

## References

ADB (Asian Development Bank). 2016a. *China, People's Republic of: Songhua River Basin Water Pollution Control and Management Project* (number 40665–012). Available at: www.adb.org/projects/40665-012/main (accessed 31 May 2016).

ADB (Asian Development Bank). 2016b. *China, People's Republic of: Songhua River Basin Water Pollution Control and Management Project* (number 40665–013). Available at: www.adb.org/projects/40665-013/main#project-overview (accessed 31 May 2016).

BBC. 2006. China plans huge river clean-up, *BBC News*, 30 March. Available at: http://news.bbc.co.uk/1/hi/world/asia-pacific/4860134.stm (accessed 1 June 2016).

Buznik, V.M. 2001. Russian Far East environmental problems, in National Research Council, Policy and Global Affairs, Office of International Affairs, Office for Central Europe and Eurasia Development, Security, and Cooperation, Committee on Improving the Effectiveness of Environmental Nongovernmental Organizations in Russia, eds, *The Role of Environmental NGOs: Russian Challenges, American Lessons – Proceedings of a Workshop*, Washington, DC: National Academy Press.

Chen, Y. 2015. Sino-Russian economic and trade cooperation in neighbouring regions and Russian's RFE strategy [Eluosi yuandong kaifa zhanlue Beijing xia de zhong'e pilin diqu jingmao hezuo tantao], *Practice in Foreign Economic Relations and Trade*, 7: 21–23.

Chu, C., Ding, L. and Huang, Q. 2014. A study on countermeasures for ecological environmental protection and management of Sino-Russian border river Heilongjiang, *Issues of Forestry Economics*, 34(3): 193–199, 204.

CIA (Central Intelligence Agency). 2016. The World Factbook: China, *Central Intelligence Agency*. Available at: www.cia.gov/library/publications/resources/the-world-factbook/geos/ch.html (accessed 12 July 2017).

Dueben, B. 2013. China–Russia relations after the Cold War: The process of institution-building and its impact on the evolution of bilateral cooperation, PhD thesis, The London School of Economics and Political Science (LSE).

FCS (Federal Customs Service of the Russian Federation). 2014. Trade Statistics. Available at: www.customs.ru/index.php?id=13858&Itemid=2095&option=com_content&view=article (accessed 1 July 2016).

Fu, Y. 2016. Are China and Russia partnering to create an axis? *Contemporary International Relations*, 4. Available at: http://valdaiclub.com/files/12062/ (accessed 12 July 2017).

Guo, L. 2013. Russia's economic development strategically orients toward East Asia and its impact to Sino-Russia regional cooperation [Eluosi jingji fazhan zhanlue dongyi de xinjucuo ji zhong'e quyu hezuo xinjiyu], *Russian Central Asian & East European Market*, 13 March. Available at: www.crc.mofcom.gov.cn/article/zhuanjialuntan/201303/86482_1.html (accessed 8 June 2016).

IEA (International Energy Agency). 2010. *Renewable Energy Essentials: Hydropower*, OECD/IEA. Available at: www.iea.org/publications/freepublications/publication/Hydropower_Essentials.pdf (accessed 12 July 2017).

Li, J. and Tian, X. 2010. Residents no longer carp about pollution in river, *China Daily*, 24 November. Available at: www.chinadaily.com.cn/cndy/2010-11/24/content_11598596.htm (accessed 8 June 2016).

Lo, B. 2008. *Axis of Convenience: Moscow, Beijing, and the New Geopolitics*, London: Royal Institute of International Affairs.

Luo, F. 1996. *An Introduction to Hydrology of Heilongjiang River Basin*, Beijing: Xueyuan Press [in Chinese].

Paik, K.-W. 2015. *Sino-Russian Gas and Oil Cooperation: Entering Into a New Era of Strategic Partnership?* OIES Paper WPM 59, Oxford: Oxford Institute for Energy Studies. Available at: www.oxfordenergy.org/wpcms/wp-content/uploads/2015/04/WPM-59.pdf (accessed 12 July 2017).

Parfitt, T. 2005. Russian town stockpiles water as industrial spill from China raises fear of deadly contamination, *Guardian* (UK), 23 December. Available at: www.theguardian.com/world/2005/dec/23/russia.tomparfitt (accessed 31 May 2016).

Sadri, H. 2014. Eurasian Economic Union (EEU): A good idea or a Russian takeover? *Rivista di studi politici internazionali*, Nuova serie, 81(4): 553–561.

Saritas, O., Proskuryakova, L. and Sivaev, S. 2015. *Russia's Water Resources 2030: Plausible Scenarios*, National Research University, Basic Research Program, Working Papers, WP BRP 42/STI/2015, Moscow.

Shen, Y. 2011. Prospect and current economic cooperation between China's Northeast provinces and Far East Russia [Dongbei Sansheng tong Eluosi Yuandong Diqu Jingji

Hezuo Xianzhuang ji Fazhan Qianjing Yanjiu], *Foreign Economic Relations & Trade*, 9: 36–39.

Shi, J. 2006. Plan unveiled to cleanse Songhua, *South China Morning Post*, 31 March. Available at: www.scmp.com/article/542704/plan-unveiled-cleanse-songhua (accessed 31 May 2016).

Simonov, E. and Dahmer, T.D., eds. 2008. *Amur-Heilong River Basin Reader*, Hong Kong: WWF.

Simonov, E., Nikitina, O., Osipov, P., Egidarev, E. and Shalikovsky, A. 2016. *We and the Amur floods: Lessons (un)learned? Report summary and conclusions*, WWF Russia. Available at: www.transrivers.org/pdf/2016floodeng.pdf (accessed 9 February 2017).

TASS. 2014. Russia's Far East authorities plan construction of new dams, *TASS*, 17 January. Available at: http://tass.ru/en/russia/715023 (accessed 8 June 2016).

Vinogradov, S. and Wouters, P. 2013. *Sino-Russian Transboundary Waters: A Legal Perspective on Cooperation*, Stockholm paper, Institute for Security and Development Policy. Available at: http://isdp.eu/content/uploads/images/stories/isdp-main-pdf/2013-vinogradov-wouters-sino-russian-transboundary-waters-legal-perspective.pdf (accessed 12 July 2017).

Voronov, B.A. 2004. Complex surveys of the Amur River basin, Amur-Okhotsk Project, First International Workshop, 3–4 March. Available at: www.chikyu.ac.jp/AMORE/2004.3kyotoSympo/03.Voronov.pdf (accessed 9 February 2017).

Wang, A. 2015. Countermeasures and suggestions of esosafety cooperation of Sino-Russian adjacent river basins, *Siberian Studies*, 42(4): 33–36.

Wang, S., Dai, C., Zhao, X. and Yang, L. 2015. Influence factor analysis of China and Russia's hydrological communication under the background of Heilongjiang mainstream flood, *Journal of Water Resources Research*, 4: 55–61.

Watts, J. 2005. China admits toxic spill is threat to city's water, *Guardian* (UK), 24 November. Available at: www.theguardian.com/guardian/2005/nov/24/frontpagenews.china (accessed 31 May 2016).

Wu, Y.-S. 2015. Russia and Chinese security, in L. Dittmer and M. Yu, eds, *Routledge Handbook of Chinese Security*, Abingdon: Routledge, pp. 90–104.

Xinhua. 2005a. China offers help to Russia in tackling river pollution, *XinhuaNet*, 1 December. Formerly available at: http://news.xinhuanet.com/english/2005-12/01/content_3865370.htm (last accessed 28 May 2016).

Xinhua. 2005b. China informs Russia of Songhua River pollution, *XinhuaNet*, 26 November. Formerly available at: http://news.xinhuanet.com/english/2005-11/26/content_3840655.htm (last accessed 28 May 2016).

Xinhua. 2013a. Chinese, Russian premiers stress cooperation on flood control, disaster relief, *XinhuaNet*, 22 August. Available at: https://web.archive.org/web/20130826141513/http://news.xinhuanet.com/english/china/2013-08/22/c_132654925.htm (accessed 12 July 2017).

Xinhua. 2013b. China–Russia friendship not 'past history': Premier, *XinhuaNet*, 22 October. Available at: www.chinadaily.com.cn/china/2013-10/22/content_17051798.htm (accessed 30 August 2017).

Xinhua. 2014. Spotlight: Energy cooperation 'backbone' of Russia–China economic ties, *Xinhua News*, 29 May. Available at: http://news.xinhuanet.com/english/2016-05/29/c_135396507.htm (accessed 30 August 2017).

Yan, J. Forthcoming. Assessment of China's transboundary water resources and water use, doctoral thesis, Chinese Academy of Sciences, Beijing.

Zhou, X., Zhang, Y. and Liu, Z. 2016. Multi-level cooperation mechanism based on collaboration among subjects of cross-border hydropower development in Heilong River basin [Heilongjiang liuyu kuajing shuidian kaifa de duocengci zhuti xietong hezuo jizhi yanjiu], *Journal of Economics of Water Resources*, 34(2): 20–40.

Zinberg, Y. 1996. The Vladivostok curve: Subnational intervention into Russo-Chinese border agreements, *Boundary and Security Bulletin*, 4(3): 76–86.

Zou, C., Ding, L. and Huang, Q. 2014. A study on countermeasures for ecological environment protection and management of Sino-Russian border river Heilongjiang, *Issues of Forestry Economics*, 3(6): 193–204.

# 9 Comparison and conclusions

## Introduction

Asian countries have varying degrees of concern about their water security. China, located on the upstream of most of Asia's international rivers, is facing increasing internal and external pressure to respond to changing demands for water. Issues including urbanization, industrial development and population growth are among the key factors that have forced the country to undergo reforms in water management. Confronted with substantial domestic water management challenges, the Chinese government is struggling to implement its water policy while also displaying respect for the neighbouring riparian states on its international rivers.

Water-sharing, in itself, is a difficult task that can generate suspicion and mistrust. From a normative perspective, countries located on the upstream of international rivers are likely to be most questioned. The UN Environment Agreement has entitled countries to exploit natural resources that fall within their national borders. This agreement also indicates that states should show responsibility in reducing environmental harm from such endeavours. However, there is ambiguity about how such responsibilities can be compatible with the exercising of sovereign rights. As water is an essential resource for human well-being, questions are raised as to whether a country's exercising of its sovereign rights will lead to failures to consider the rights and well-being of downstream states. Naturally, China, as a non-member to global water norms, is particularly questioned about its position as an upstream state.

In this chapter, we investigate China's interactive behaviour developed as a result of its sharing rivers with neighbouring riparian states. The empirical findings from four case studies provide rich material to understand the rationale in China's diplomatic initiatives taken with its riparian neighbours. In the four cases, the riparian states have shown varying responses to China's appropriation of water resources. Although most are faced with alarming water security concerns, mixed responses have been shown that do not solely constitute concerns over water management. As individual actors' interests vary, China's water diplomacy has resulted in different cooperative initiatives, be they water-related or non-water-related activities. By comparing the cases, we investigate contextual factors impacting on the rationale and outcomes of China's water diplomacy. Attention is placed on understanding how China has developed diplomatic initiatives that reflect a combination of varying national interests in water resource management as well as the state of its foreign relations.

This chapter is structured in three further main sections. In the next section, the results of the case studies will be compared and analysed with respect to the rationale of China's water diplomacy, which reflects the dynamic interaction of various interests. Part of this comparative analysis is to investigate how different interests that impact on China's development of water diplomacy have been articulated. In the subsequent section, the research questions, as formed in the first and the fourth chapters, are answered and conclusions drawn. Our analysis focuses on the contextual factors upon which China develops water diplomacy, especially when water security concerns exist. These will help us to address the key research questions set out in the introduction. The final section examines China's perception of water as a security threat and the country's subsequent policy responses.

## Comparative analysis of the case studies

### A *methodological introduction*

The investigation of China's water diplomacy in the four case studies has illustrated that it reflects various factors relating to the planning and management of specific rivers as well as concerns for foreign relations. This study has chosen four river basins as case studies to provide sufficient diversity with respect to the GDP per capita, level of water security, challenges to trust-building, and regime type of the countries within each basin. Most of the countries represent a deficient democracy in their political regimes. Before analysing and comparing how China develops diplomatic initiatives with its riparian neighbours in the four river basins, we briefly reflect on the contextual diversity of the case studies.

The Ganges–Brahmaputra–Meghna (GBM) river basin originates in the Himalayan Mountains in China, and runs through Bhutan, India and Bangladesh. Sino-Indian water sharing relations are selected as a sub-case. India has developed a troubled relationship with China. Being one of the world's largest developing states, it is faced with an increasing water demand relating to its rapid industrial development, escalating population growth and increasing agricultural needs in an era of climate change (Cronin *et al.*, 2014). Like China, the country is considered to have a low level of water insecurity over the Yarlung Zangbo River, although it has displayed a high level of water insecurity on the Ganges and the Indus Rivers. Its local-level GDP per capita is particularly low in locations along the GBM that border with China. In the democratic society of India, various local actors have displayed disagreement with China's upstream appropriation of the GBM's water resources. While the public has held a deep historical mistrust towards the Chinese government, the different administrations have shown varying stances. The previous Singh government was calm towards China's exploitation of the GBM's water resources. However, the current Modi government has shown an intention to politicize water issues, hence showing an intensified disapproval of the Chinese government's exploitative activities.

The Mekong River basin originates in Yunnan Province in south-west China and traverses another five riparian countries of the Greater Mekong Subregion, comprising Vietnam, Cambodia, Laos, Thailand and Myanmar. These countries are also called Lower Mekong Countries (LMCs) and they share with China a relatively low level of water insecurity on the Mekong River, although they show varying concerns in each country. Most of these countries are observed to experience a weak liberal democracy as strong support for authoritarianism remains (Springer, 2009). Politically, they are strategic collaborators, mostly through the Association of Southeast Asian Nations (ASEAN)[1] which is one of the major coalitions in the region. China has shown continuous economic and trade-related cooperation with the ASEAN countries while developing its strategic interests in the region. Political tensions and border disputes have been intertwined issues in China's relations with some of the LMCs, hence China's water cooperation with these nations is likely to result in some level of distrust.

LMCs are displaying their strong concerns about the impact of China's damming project on the Mekong River. Overall, the concern and disapproval involves both water- and non-water-related issues but contradictory stances have been shown towards China. As each riparian state has different geographical locations on the Mekong River basin, their incentives to collaborate with China also vary. Since there is a lack of basin-wide coordination, interstate and intra-state conflicts have also arisen. Apart from country-specific water management concerns, the LMCs have also expressed their combined concern about China's pursuit of its strategic interests in the region. Complicated disapproval of China's activities therefore emanates from this region.

Shared between China and Kazakhstan, the Ili and Irtysh Rivers are of high importance to central Asia. Both rivers are strategically vital waterways for Kazakhstan. Like China, Kazakhstan has a high level of water insecurity on the Ili River and it is now showing a strong and increasingly alarming level of concern towards water management. Its GDP per capita reaches over US$5,000. The country has initiated a democratic political system in name, but in reality its performance rates as a deficient democracy (Sørensen, 2010). With direct competition for water existing with China, both the Kazakh government and the public have focused on the allocation of water quantity between the two countries. For China, the river basin is of great importance for realizing the stability and prosperity of Xinjiang Province, which is geographically located at the core of the region. The Chinese government has also demonstrated its strategic interests in promoting Xinjiang's development, which is seen as essential to link the European and Asian continents (Wang and Jia, 2016). China and Kazakhstan have developed relatively stable relationships in various fields, including through strategic economic, trade and defence partnerships. Nevertheless, with a high level of competition over water quantity, Kazakhstan has shown continuous concerns over the shared Ili and Irtysh River basins.

China has developed institutional cooperation with Kazakhstan to facilitate collaborative water management. State-led cooperation has been established, starting with meteorological and hydrological cooperation for forecasting weather and water-related disasters, and gradually shifting to more substantial water-management cooperation involving collective monitoring activities and scientific research. Various Kazakh authorities have worked with China, including with its Ministry of Environment and the Ministry of Agriculture. With the Kazakh ministries, both countries have decided to cooperate in technological development, information-sharing and most recently the construction of a project that allocates water between the two countries.

China shares the Amur River basin with Mongolia, Russia and North Korea. The Amur River runs eastwards from Mongolia through the north-east region of China. China has shown a low level of water security over this river basin. It consumes a high volume of Amur River-sourced water for agricultural uses, while hydroelectricity development and transport also account for a growing proportion of the consumption. In comparison, Russia enjoys a low level of water insecurity from the Amur River basin, contrasting with a high level of water insecurity that China experiences on the river. Residents along the Amur River in the RFE (Russian Far East) enjoy a high level of GDP per capita of over US$2,700. The Russian public have expressed concern over water pollution and disasters that first occurred on China's part. Since the Russian government has kept relatively calm, disapproval in Russia was well contained. Apart from these areas of concern, Russian environmentalists have also expressed concerns about the construction of dams, although this was not necessarily directed towards China.

Russia and China have experienced historical burdens involving political and ideological disparities. They have developed relatively stable relations although in ambiguous and uncertain ways. With the occurrence of water pollution incidents in recent years, the two countries have reached agreements on how to best collaborate to alleviate such disastrous events. Institutional cooperation for water management has been formed, including for collaborative responses to flooding and emergency events. However, China's water diplomacy with Russia has been complicated by the differences between its own and Russia's strategic interests over the Amur River, which is crucial to both countries' regional strategic interests in East Asia. China aims to promote regional development that is part of its Eastern Silk Road Economic Belt strategy, while Russia aims to promote the development of the RFE. China's water diplomacy that is characterized by prioritized strategic interest in the RFE has the tendency to impede its water diplomacy.

### *Institutional cooperation in China's water diplomacy*

Using the theoretical framework developed in Chapter 4, water management institutions (WMIs) refer to institutions that result from cooperative attempts among riparian states. In the four cases, varying levels of diplomatic cooperation has led to different WMIs being developed between China and its major riparian neighbours. With regard to the level of cooperation, Sino-Indian cooperation has formed only a relatively low level of WMIs. In comparison, Sino-Russian cooperation has become more substantive, and slightly more so than that between China and Kazakhstan. The number of agreements formed between China and Russia also exceeds those between China and Kazakhstan. China's highest level institutionalized cooperation exists for the Mekong River. This is demonstrated not only in the level of formality of state-led cooperation, but also in the scope of collaborative water management between China and the Lower Mekong Countries. Accordingly, its WMIs and length of cooperation with neighbouring riparian states also vary, ranging from more than three decades (with Russia and Kazakhstan), two decades (with India), and less than two decades (with Lower Mekong Countries). In this section, the level of institutionalized cooperation is examined from the following aspects: forms of diplomatic cooperation, the inclusiveness of participants and the scope and area of interests.

### *Forms of diplomatic cooperation*

In most countries, states act as diplomatic representatives to deal with issues surrounding transboundary water management. Similarly, the Beijing government has also played this role. In most cases, political authorities have sought opportunities to develop institutional mechanisms that best suit the country's interests. Bilateral mechanisms remain the dominant form of cooperation. This is seen in three cases – in China's sharing of the GBM, the Ili and the Amur River basins where cooperation has been developed with India, Kazakhstan and Russia, respectively. Only in the case of the Mekong River has a multilateral mechanism been developed. In this case, China has participated in the Greater Mekong Subregion Mechanism (GMSM) initiative to develop the multilateral Mekong Cooperation Mechanism (LMCM).

China has formed different levels of diplomatic cooperation with its riparian neighbours (see Table 9.1). In all four cases, China has established state-level cooperation with the governments of its neighbouring riparian states and this usually involves institutional initiatives between concerning authorities. The Ministry of Water Management and the Ministry of Environmental Protection are the two Chinese authorities mostly involved. Such cooperation has also taken the form of establishing a special commission in charge of the implementation of water cooperation agreements (see Table 9.2). Joint committees and working groups have been established at the ministerial level to ensure implementation. Furthermore, it is also noted that project-specific institutions have also been established to facilitate deepened collaboration. In the case of the Ili and Irtysh Rivers, both China and Kazakhstan have agreed to the joint construction of the Khorgos River Friendship Joint Diversion Control Project of 2011 and a special committee had been established to implement the project.

*Table 9.1* Level of water diplomacy

| | *India* | *LMCs* | *Kazakhstan* | *Russia* |
|---|---|---|---|---|
| Forms of agreements | MOUs | MOUs and water agreements | MOUs and water agreements | Water agreements |
| Number of agreements | 5 bilateral MOUs | 5 multilateral agreements | 11 bilateral agreements and MOUs | 17 bilateral agreements and MOUs |
| Scope and areas of interests | Disaster control | Disaster control, quality control, biodiversity protection, navigation, power generation | Biodiversity protection, quality control, disaster prevention, power generation | Biodiversity protection, disaster prevention, quality control, navigation, power generation |
| Level of institutionalization | Low | Very high | Medium | High |

*Table 9.2* Institutional unit responsible for water cooperation

| | *Institutions in charge* | *Responsibilities* |
|---|---|---|
| India | Joint Rivers Commission | Data exchange on flooding aversion |
| Kazakhstan | Joint Commission on the Protection of the Environment; China–Kazakhstan Cooperation Commission; Special Committee of the Horgos Project | Data exchange, and the joint construction of the Horgos River Friendship Joint Diversion Control Project |
| Lower Mekong Countries | The Lancang–Mekong Cooperation Mechanism | Data provision by China, facilitate cooperative activities |
| Russia | Ministry of Water Resources and relevant Russian ministries | Data exchange between the two countries |

Only in two cases, regarding the sharing of the Mekong and the Amur Rivers, does China have an interest in incorporating local governments in the development of collaborative water management initiatives.

### *Inclusiveness of participants in WMIs*

As mentioned in Chapter 3, national authorities have played a dominant role in the management of water resources. The formal diplomatic cooperation between China and its neighbouring riparian states has indicated that institutional arrangements have been established to reinforce the water management authority of specific Chinese government departments such as the Ministry of Water Resources and the Ministry of Environmental Protection.

As identified in the literature, most developing countries are limited in their ability to establish integrated water-resource management initiatives. These countries lack the participation of key stakeholders in basin-scale management. Non-state actors have been found to have a weak capacity to play a significant role in the policy processes of water governance or to participate in water management practices. However, as shown in the empirical findings, in countries where national NGOs have been active, multiple stakeholders have had impacts on China's planning and management of its shared rivers. In the case of the Mekong cooperation, pressure from anti-dam non-state actors resulted in Chinese authorities incorporating environmental impact assessments into the planning for dam construction projects. Although procedurally such acts were incorporated after the project had started, it has proven that NGOs have become significant stakeholders in the monitoring of state-led cooperation. With the existing GMSM initiative, the Chinese government has also participated in biodiversity protection that incorporates local government, local communities and NGO collaboration. Similarly, in the case of China's water diplomacy over the Amur River, enhanced cooperation has been achieved through the collaboration of local governments from both countries. Such arrangements indicate that a certain degree of decentralization has been achieved in China's collaborative water management initiatives.

### *Areas of interest in collaborative water management*

In China's diplomatic initiatives for the collaborative management of shared water resources, there are three identified areas of particular interest – disaster prevention, quality control and biodiversity protection. A strong focus is also placed on power generation. Apart from India, activities have been developed in these areas of interest in all four of the case-study river basins (see Table 9.3).

A primary collaboration concern is to develop disaster prevention and management systems that respond to water-related emergencies such as flooding and water pollution that may result from industrial discharges. The prevention of disasters primarily focuses on flooding. However, in most cases, only superficial collaboration has so far been reached. Collective actions have been developed, but only passively, and so the ability to prevent disasters is yet to be tested. For instance, Sino-Russian collaboration has been reached to conduct collective water monitoring activities, but there has been no follow-on work developed. With Kazakhstan, concerns about water quality has seen China strengthen its quality control measures. This collaboration reflects an extension of a key policy aspect of China's domestic water management concerns. In contrast, for the Mekong River, measures of quality control have barely considered the prevention of pollution from navigation or dam construction projects.

Interestingly, water scarcity is not perceived the same as flooding by China. In the case of the shared Ili and Irtysh River basins, there is competition for water resources between China and Kazakhstan. China has not yet developed water diplomacy as promptly and closely on the issue of water scarcity as has done for flooding prevention. This may be because water scarcity has not yet reached an alarming level for the country to feel threatened. At present, water scarcity remains a challenge, but it is seen to be a manageable challenge in domestic China.

Water quality control is another concern shared by neighbouring riparian states. Industrial development in China has posed threats to regulating the water quality of its shared rivers. Therefore, collaboration is mostly undertaken to prevent industrial pollution, as seen for the Ili and Irtysh, the Mekong River and the Amur River basins. However, in places where there is little sign of pollution, such as in the sharing of the GBM, there has been little incorporation of concerns about industrial pollution prevention into the collaboration. This is probably because the Yarlung Zangbo River, the part that runs through China's territory, is located in remote areas and hence unlikely to become polluted.

Biodiversity protection is another area that China has begun to collaborate on with its riparian neighbours. This policy primarily addresses the loss of

*Table 9.3* Level of cooperation China's water diplomacy

| | *India over GBM* | *LMCs over Mekong* | *Kazakhstan over Ili and Irtysh Rivers* | *Russia over Amur River* |
|---|---|---|---|---|
| Collective programme | Disaster alleviation | Disaster alleviation, biodiversity protection, navigation, power generation | Disaster alleviation, biodiversity protection, navigation, power generation | Biodiversity, disaster control, navigation, power generation |
| Non-water issues | Border dispute | Terrorism, safety | Other political issues, e.g. 'Three Evils' of separatism, extremism and terrorism | Border dispute |
| Institutional cooperation | Low | High | High | Relatively high |

biodiversity in river basins. For the Amur and Mekong Rivers, ecological protection has been incorporated into the water diplomacy. This generally started as a direct response to the concerns of China's neighbouring riparian states. Cooperation for biodiversity protection between Russia and China has occurred since the 1980s but it was not until very recently that more collaborative programmes have been developed. Similarly, for the Mekong River, China was a participating member of the GMSM's biodiversity protection programme. Ad hoc cooperation that includes collective research activities has also occurred to support the protection of the river's biodiversity values. Consequently, both countries have been committed to collectively monitoring water quality and the preserving suitable water ecology. Such work has since been expanded to at least two threatened native fish species. This work has much potential to continue in the future.

To sum up, different water- and non-water-related concerns have been incorporated into the collaboration between China and its neighbouring riparian states. The responses from the neighbouring states have reflected not only their level of trust, which is impacted by foreign relations, but also their domestic water management capacity. The findings confirm previous research that water quality is one of the key concerns that countries are likely to disagree on when water conflicts occur (Wolf *et al.*, 2005). Apart from Kazakhstan, in most cases, since little competition exists between China and its neighbouring states over water quantity, China's water diplomacy has proved to accommodate different interests through water agreements. However, acute competition exists between China and Kazakhstan over shared river resources, China's diplomacy has not been smoothly developed with Kazakhstan. The key disagreement of the division of water quantity has barely been touched upon by the two countries.

### Factors influencing China's water diplomacy

China's cooperative behaviour can also be placed in the context where it interacts with different riparian states that have shown vested interests in developing diplomatic collaboration with China over shared rivers. Responses from these other countries represent a mixture of complicated water management and foreign policies. In this section, we assess the influence of contextual factors on the extent and features of China's water diplomacy. Factors include neighbouring countries' asymmetries in security over water resource use, the nature of their political regimes and associated challenges to trust building, and the economic development level (see Table 9.4).

What can be learned from the four cases is that the degree of institutionalized water diplomacy formed has a positive correlation with countries' asymmetries in water security. This is particularly the case when China is threatened in the sharing of its international rivers that are between China and Kazakhstan and Russia.[2] In situations where smaller asymmetries exist such as with Kazakhstan in the Ili and Irtysh Rivers, diplomatic cooperation is formed to a higher degree than that formed with Russia on the Amur River. In these cases,

*Table 9.4* Contextual factors in the four cases

| | *India over the GBM River* | *LMCs over the Mekong River* | *Kazakhstan over the Ili and Irtysh Rivers* | *Russia over the Amur River* |
|---|---|---|---|---|
| Asymmetries in water insecurity | Low | Low | Low | High |
| Political regime | Democracy | Deficient | Deficient democracy | Deficient democracy |
| GDP per capita | Very low | Low | High | High |

threats of water scarcity and limited use of water resources from river basins have alerted the Chinese government to develop water cooperation with riparian neighbours. The findings bring support to previous findings on the causal link between levels of asymmetries in states' security over river basins and the likelihood that they may succeed differently in the forming of collaborative water resource management (Ovodenko, 2014).

It is also noted that the country's incentive to form water diplomacy is not immediately affected by its water security level. In situations where China enjoys a high level of water security, as in the shared Mekong River, the country is more likely to succeed in forming water diplomacy with riparian neighbours that share similar level of water insecurity. However, in Sino-Indian water cooperation, a high degree of asymmetry exists and it appears that negotiation on water cooperation becomes harder.

Additionally, China has shown equal interest to form water diplomacy in situations when it is not threatened. We can infer that China's behaviour in water cooperation is not strongly affected by the level of water insecurity it is faced with. In the case of the sharing of the Mekong River, China and the LMCs enjoy a high level of water security and the Chinese government has shown strong commitment to conduct peacemaking activities to secure the safety of citizens across its borders. This reminds us that the rationale for the Chinese government to develop diplomatic cooperation is complex. To Chinese policymakers, scientific findings on water-related disasters does not constitute a determining factor that drives the country to form water diplomacy.

When focusing on the nature of political regimes, it is noted that most of the countries are democratically deficient states. No strong democracy is found in the chosen cases. There are no clear causal relations between the robustness of the neighbouring country's democratic system and the success of water diplomacy that China has developed with that country. In scholarship, it is suggested that democratic systems are likely to show a higher level of commitment to global norms (Brochmann and Hensel, 2009; Gerlak and Grant, 2009). Findings from this study do not provide direct support for this conclusion. In this study, most countries display a weak democracy, hence they lack the ability of robust democratic societies in effectively dealing with situations associated with the management of transboundary rivers. For example, they typically lack the ability to clearly articulate and successfully negotiate their political and environmental interests at multiple levels.

When focusing on the economic development level, it seems that China is more likely to succeed in developing a higher level of institutionalized water diplomacy with richer states. There is a correlation in countries' economic development and their responses to China's cooperative water management initiatives. Richer societies are likely to experience an advanced level of ecological modernization, hence better capacity in the management of water resources. Kazakhstan and Russia are two relatively well-off states, in which national policy documents reveal concerns about holistic water management, and China has developed better water management cooperation with these states. However, less has been developed with India, whose GBM river-basin communities suffer from low GDP per capita. Hence, it is less likely that they will develop a suitable water-sharing mechanism. This may indicate that poor countries are likely to experience difficulties in domestic water management and be less able to respond efficiently to water governance challenges. An exception lies with the LMCs, with which China has developed institutional cooperation. This may result from the fact that the Mekong River basin has received international attention. The LMCs have been strongly influenced by international actors with regard to the importance of basin-wide management, hence they have shown willingness to commit to state-led cooperation.

## China's rationale to develop diplomatic initiatives

In most cases, China has shown flexibility in its water diplomacy and variation in its cooperative initiatives where a combination of different interests exists. Trust-building and peacemaking are commonly found in China's water diplomacy. The four cases show variation in how China combines these identified interests in its water diplomacy. In some cases, certain links have been built between the identified interests but the links between the diplomatic initiatives are not always clear.

### *Reduce mistrust with riparian neighbours*

Trust-building refers to initiatives that China has taken in interacting with its riparian neighbours to implement its water policies. Initiatives of this type aim to reduce the mistrust that may occur between China and its riparian neighbours, particularly when foreign relations are strained. Principally, China has intended to exercise its sovereign rights and appropriate water resources within its territory. However, as an upstream state, China considers that its rights have not been protected well by existing global water institutions and the country has therefore refused to prescribe to them. However, the national government claims that it upholds the norms such as 'no harm' and prior consultation with downstream states. Easily found in the four river basin case studies is China's intention to reduce the mistrust which may arise from its unilateral upstream actions. Indeed, this is something that China seeks to achieve as a first stage of cooperation. This displays China's intention to listen and to respect others in its water diplomacy.

China has responded to disagreements with its neighbouring riparian states by showing an intention to share water data. This demonstrates that it has rarely developed the intention to use water data as a military tool to constrain downstream states. Subsequently, the measures that the Chinese national government has adopted, in most cases, are to share data that has been collected from hydro-stations to monitor water changes associated with upstream constructions. Naturally, there is conflict between the water data that China is entitled to keep for domestic purposes and the level of data transparency that could avoid its downstream neighbours feeling threatened due to the asymmetric possession of information.

### Peacemaking and the protection of human security

Peacemaking refers to actions to protect human security that may be endangered by the sharing of international rivers. China's peacemaking intentions are evident in all four cases, where it demonstrates a commitment to reducing the negative effects of natural disasters or those that may result from its dam construction projects.

Peacemaking initiatives can be guaranteed by the effective implementation of sharing or exchanging information. Underpinning China's peacemaking initiatives is the security of its own citizens in the management of shared river basins. The rationale primarily relates to China's advantaged upstream geographical location, from which China sees itself as obligated to prevent disaster events in downstream states. With asymmetric possession of information, China understands that downstream countries are in a disadvantaged position to effectively manage their water resources. More specifically, flood control is one of the identified issues that can have a large impact on human security. China feels almost obliged to work with its riparian neighbours, as strong suspicion exists that flooding disasters are caused directly or indirectly by China's upstream damming activities. Such actions have sometimes been taken even before systematic diplomatic cooperation had occurred. For instance, in the case of the sharing of the GBM, China had attempted to inform downstream countries about foreseeable disasters as early as 2000. Similarly, China has intended to avoid downstream flooding events on the shared Mekong River. In all four cases, China is therefore committed to reducing the negative effects that its upstream activities may produce, with emergency control being the key area of concern; and strong emphasis is placed on addressing this issue.

### Water resource management

In some instances, China considers that better profits can be achieved by developing collaborative water management with its riparian neighbours. A certain degree of collaboration has been developed for both China and its riparian neighbours to benefit from water-sharing while also addressing common water management concerns.

Many downstream countries are incapable of foreseeing water-related natural disasters. As a result of their lack of appropriate facilities and methods, these countries have been weakly equipped to protect their citizens against such disasters. Hence, a primary collaborative water management step for these countries is to establish technological cooperation and collaborative research with China. China has also facilitated water management improvements in both of these areas, such as through the development of irrigation projects and securing adequate water flows for ecosystem services. These usually are multifaceted activities that include the prevention and management of pollution, emergency management and ecological protection. In cases where substantial collaboration has been built, the empirical cases indicate that China also has interests in developing power generation and tourism projects with its riparian neighbours.

In sum, peacemaking is a fundamental incentive for China to develop its water diplomacy. A key concern identified in all four cases is that China has shown no distinction between its own citizens or those of its downstream neighbours. Developing trust-building and collaborative water-management activities are closely linked intentions. However, negotiations over developing collaborative water management have not been a linear process. Despite that China may combine concerns over water resource management and political interests, depending on the contextual factors, its water diplomacy may have only a limited impact on the development of sustainable water basin management. In fact, for China, the purpose of its water diplomacy may well have a single focused and it does not necessarily incorporate the improvement of water management. This indicates that a feature of water diplomacy in Asia is great formality that does not necessarily result in better water management. Therefore, unlike the presumptions of institutionalism, in the Chinese case the formality of water diplomacy does not necessarily lead to a high level of sustainable development. This explains why in some cases, such as with India, although diplomatic initiatives have developed successful peacemaking activities, these did not necessarily result in China developing collaborative water management with India.

## Answers to research questions

The four cases show the evolving nature of China's efforts to resolve water-sharing conflicts. Across different river basins, China has shown different vested interests in its water diplomacy. This section answers to the research questions posed in Chapter 1.

### A *To what extent is China likely to adopt global water norms?*

In most cases, China's unilateral appropriation of water resources has been challenged but the country has not signed any agreements to conduct prior consultation. As learned from the four case studies, the Chinese government has shown a certain degree of recognition of global water norms. To some extent, some of the norms have been complied with. However, the Chinese government has prioritized its national interests in the planning and management of the shared water basins. In most cases, it has managed to push through its water policies but in rare cases it has had to compromise its water construction projects. Among all adaptive measures taken in China's recognition of global norms, the principles of 'no harm', 'fairness' and 'benefit sharing' are the main principles that have been adopted in China's water diplomacy.

'No harm' is a principle highlighting that a country's planning and management of water resources should respect the interests of other riparian states, and that consultation should be adopted to ensure that the interests of other countries are taken into consideration. However, China primarily resorts to unilateral actions for the planning and management of its shared rivers. The Chinese government advocates that it has adopted various approaches to ensure that the interests of lower riparian states are considered in their appropriation of water from transboundary rivers. One approach in China's compliance with the 'no-harm' principle is to share hydrological data with its riparian neighbour states. The Chinese authorities intend to demonstrate that no serious harm had been made to the downstream states. For example, the government stresses that it has been adjusting its policymaking processes to comply with scientific decision-making in environmental projects. In the Mekong River, local communities and environmentalists have demanded the disclosure of environmental assessment information and that the assessment process be made more transparent. After doing so, the outcome for China was not positive, with Myitsone Dam Project having to be halted (see Chapter 5).

It should also be noted that the Chinese authorities have placed keen attention on certain areas of water management, particularly when physical harm or threats are immediately seen, such as the possibility of flooding or water scarcity. Ecological impacts are of lesser concern to the Chinese government in their management of the shared river basins. Hence, China primarily focuses on the prevention of physical threats posed to humans, including through flooding, drought and water pollution (He, 2008).

Equity in water-sharing is another underlying concern in the 'no-harm' principle. One of the implied rules in this multiple-meaning principle is that countries share profits in equitable ways (McIntyre, 2014). In several cases, such as with the LMCs and Russia, benefit sharing has been seen in China's exploration of economic development opportunities with its riparian neighbours. In these projects, both China and its neighbours have decided to appropriate the shared water resources, such as through dam and navigational construction projects. In the case of Kazakhstan, with water volume being the largest constraint between the two countries, only through sharing benefits from a common project has China managed to contain Kazakhstan's increasing concern over the security of water resources from the shared rivers. The Khorgos River project represents a successful initiative whereby both countries agree to share equally the rivers' waters for irrigation projects. The project is also of significance for controlling flooding as well as promoting trade, which demonstrates a level of strategic economic collaboration between the two countries. China believes that the Khorgos River project demonstrates how benefit sharing could ease concerns over shared waters. From this perspective, sharing benefits seems a shortcut for China to persuade downstream riparian states to accept its proposals in the exploitation of water resources.

Current global water norms have only provided broad rules in promoting equity and fairness in water-sharing. China's water diplomacy shows that the country has paid attention to this principle; however, it has not revealed its own interpretation. In the case of water-sharing in the north-west where particular concerns over water quantity exist, China has attempted to promote a fair approach to its water-sharing with Kazakhstan. Additionally, its understanding of fairness is rather narrowly developed, focusing on the timing and ability to access water resources. A 'first serve, first use principle' exists, which emphasizes the timing of water users in their appropriation of water resources. Those who first appropriate water are granted the rights to that water. Similarly, those who are closest to the water source are advocated to be endowed with the right to acquire that water (Li *et al.*, 2011). As seen in the case of sharing the Ili and Irtysh Rivers, such principles have raised disagreements from the Kazakh side. As a result, the negotiations between China and Kazakhstan have not yet reached satisfying outcomes.

As a powerful actor located upstream on numerous transboundary rivers, it can be inferred that China's intentions have begun to display a norm-oriented power construction path (Pu, 2012). Being aware of the spillover effects of its appropriation of water resources within its territory (Feng *et al.*, 2015), China has made gradual changes in its foreign policy agenda for dealing with water disputes. Although China's upstream location physically facilitates the country to do so, its behaviours are constrained. Considering the negative effects that unilateral upstream actions may bring, China has shown caution in its appropriation of water resources.

### B *What are the outcomes of China's diplomatic efforts to resolve conflicts and promote sustainable water management?*

Faced with pressure from environmentalists and non-state actors in riparian countries, such as the Lower Mekong Countries, China has given greater attention to promoting integrated water management. China's diplomatic cooperation has resulted in various forms of sustainable water management. In other words, although China's diplomatic cooperation over water primarily focuses on securing its domestic economic and political interests, it is also likely to result in collaborative water management with the country's neighbouring riparian states.

China has started to formalize water management and develop collaborative water management projects. Viewed from the extent to which water diplomacy has been developed, China's water cooperation has, to varying degrees, incorporated the agenda of sustainable river basin development, which is based on substantive collaborative water management. As developed in Chapter 4, two interlinked dimensions are identified regarding China's behaviour over shared water resources (Table 9.5). These are the extent to which collaborative water management has been achieved (vertical level), and the level of institutional cooperation that has been formed between China and its riparian states (horizontal level). There are three distinct situations where China's water diplomacy has succeeded (see Table 9.5). Scenario A indicates that a low level of institutionalization has seen few initiatives achieved for sustainable development. Scenario B represents a situation where a certain degree of institutionalization

*Table* 9.5 Scenarios of China's behaviour and its implications for promoting basin-wide diplomacy

| | *Low level of institutionalization* | *High level of institutionalization* |
|---|---|---|
| Low level of cooperation in water-basin collaboration | A | B |
| High level of cooperation in water-basin collaboration | C | D |

has been achieved, however, few sustainable development initiatives have been achieved for the river basin. Scenario C does not apply in reality. Scenario D indicates that a higher level of formality has also resulted in positive outcomes for sustainable water basin management. The four cases represent distinctive situations.

Scenario A, represented by China's sharing of the GBM with India, has proven to be the worst situation among the case studies, with only a low level of sustainability initiatives being developed. In this case, China's diplomatic initiatives represent a symbolic gesture that takes little consideration of water resource management. Such WMIs may potentially endanger China's foreign relations with its neighbours, and may lead to political tensions. Essentially, both China and its neighbouring riparian state(s) (in this case India and Bangladesh) benefit little from such water cooperation.

With Russia and Kazakhstan, China's water cooperation reflects scenario B where a higher level of institutionalization has been achieved. However, little has been achieved in sustainable water management. In both cases, water diplomacy has extended from the typical issues of water quality control, disaster prevention and power generation to incorporate new areas of concern such as biodiversity protection. It is also noted that in both cases, China has implemented decentralization in its water management to some extent.

The Mekong River case represents scenario D, which is the best scenario for China as well as for all other nations involved. China takes a proactive position in promoting sustainable water resource development at a regional scale. China has developed a broader consideration of basin-wide interests while successfully developing institutional agreements with other nations. Being aware of the impacts of water-sharing disagreements for regional security, China has shown a firm commitment to developing a high level of institutional cooperation with the Lower Mekong Countries. However, in comparison to scenario A, this situation probably also proves to be the most challenging of China's basin-wide management initiatives. With the LMCM that China has developed, the country is committed to providing financial and technical support to all relevant riparian states of the Mekong River basin. By doing so, the country has taken a leading role in coordinating the management of this river basin.

The rationale for China's water diplomacy varies and its impacts on sustainable river basin management also vary. When viewing the three scenarios, they represent distinct situations where widely varying dynamics exist between China and its riparian states. There is no sign that a linear path exists from scenario A to scenarios B or D. Scenario A does not necessarily evolve into scenario B, as its focus may remain on facilitating trust-building. Naturally, such an intention is unlikely to lead to substantive collaboration for a river basin's sustainable development.

Similarly, scenario D should not be been seen as a linear development from scenario B. In the case of scenario B, water diplomacy indicates dynamics where a certain degree of collaborative water management has been developed as a result of interactions with riparian neighbours, such as in the case of Kazakhstan and Russia. In both cases, China's water diplomacy has somehow temporarily succeeded in peacefully resolving disagreements with Kazakhstan and Russia without developing substantial basin-wide sustainable development initiatives. However, there are potential tensions between China and the other two countries. For instance, Kazakhstan's concerns over the shared water resources have not been sufficiently addressed. With Russia, China's placing emphasis on realizing strategic interests may also raise mistrust. At the same time, both Kazakhstan and Russia have shown interest in conducting collaborative water management. Where this driving factor exists for China to incorporate the concern in its agenda, scenario B may upgrade towards D. Again, when scenarios A and B are compared we are reminded that the contextual factors are important for such a transition to take place, as scenario B enjoys higher economic development level.

## C *What is China's rationale in forming water cooperation?*

Water is a multifaceted resource for China. Understanding China's rationale for water diplomacy essentially is to investigate China's varying perception of water insecurity in water sharing. As analysed above, the nature of China's water diplomacy clearly differs in the four cases. These case studies indicate that China has continued to use peaceful measures in pursuing its interests when interacting with other riparian states, although it has displayed varying cooperative behaviours, dependent on the differing interactions with these states. As national interests are the primary concern in China's forming cooperative measures, we see that realism remains a useful theoretical approach to explain China's rationale and responses to situations such as sharing transboundary rivers. Water, as a multiple-natured issue, has created complexity in China's security agenda.

### *Water as conventional security concern*

China has shown a realistic approach that prioritizes its domestic interests over those of other countries, with both its economic and defence values being of particular interest. Water is primarily viewed as a source of material profits that are core to China's national interests. In all cases, water represents a natural resource that is crucial to the country's economic development, such as through the construction of navigational or hydropower projects. Such economic development opportunities are reflected in China's existing water-related policies for shared river basins. More specifically, China views water as an important source of profits. In most cases, water diplomacy facilitates China's strategic interests nationally and internationally. Apart from India where there are sensitivities associated with the shared rivers in South Tibet, in the rest of the cases China has strongly relied on the shared river basins for regional development, such as Xinjiang in the north-west, Yunnan and Guangxi in the south-west, and Heilongjiang in the north-east. With transport developed along these river basins, the boosted regional development has also been extended to promote the 'One Belt, One Road' strategy designed to expand China's global economic and political influence. This indicates that the country has prioritized a regional development strategy that relies on water's economic values. From this perspective, China has shown a territorial concern in appropriating water resources that are located within its national borders.

At the same time, China as the upstream country has gained a natural geographical advantage over downstream countries as information asymmetries exist between China and other riparian neighbours. Defence interests, at times core to China's national interests in its accessing transboundary water resources, may be vested in China's securing of economic interests from water, such as in its water diplomacy with India and Russia. In the case of India, China's negotiation over water issues is combined with that on border disputes and military cooperation. In the case of the Amur River, China's cooperation with Russia has resulted in it gaining Heixiazi Island – an island that the two countries have historically disputed. In both cases, water diplomacy has somehow facilitated positive outcomes for long-held border disputes, further indicating that the Chinese government has succeeded in its foreign affairs initiatives.

The nature of China's water diplomacy clearly differs in the four cases, although developing sustainable water management initiatives has not been the primary concern in China's water cooperation in all four. It should be noted that China's water diplomacy represents an enhanced commitment in foreign policy for it to proceed with generating profits from its water resources and does not necessarily indicate that the Chinese government will prioritize the development of substantive water sharing collaboration.

### *Water as a security concern*

Water, and the potential environmental changes it may bring, has been viewed as posing threats to national interests, resulting in varying associated behaviours. The empirical findings in this study indicate that the Chinese government perceives water as a potential threat to national security, either as a threat to physical safety or to the country's economic interests. In situations where heightened competition exists between China and its neighbouring riparian states, China has shown different responses. And China's consideration of the environment as a political issue depends on various factors. Unlike in Western Europe societies, where scientific findings on the scale and scope of environmental change are a determining factor for how such issues are perceived, this may not be the case in China. To Chinese authorities, the nature of environmental change alone is not the sole factor that affects their adopting environmental issues in their security agenda. As discussed in Chapters 1 and 4, security studies have highlighted various concepts of environmental security that haves been heatedly debated in Western societies. However, we should be cautious when applying these concepts to China, where the perception of water is multifaceted. Two contrasted views can be identified.

1 Water insecurity, or water-related disaster, is most often perceived as a non-security issue. The Chinese authorities have developed a decisive attitude to the natural world (Shapiro, 2012). Rooted in its national identity, the Chinese take a technocratic view towards the environment, and have displayed confidence and familiarity towards water management. Vulnerability around water issues is rarely perceived as a series of extant threats, as it is often assumed that water security is kept under control. Such responses are also based on Chinese authorities' confidence in efficient policy implementation. It is believed that the Chinese authorities have been flexible, and ready to adjust human uses of natural resources to create harmony between the needs of human society and nature (Zou, 2014). This explains why initiatives for involving the public have been developed rather slowly in China's environmental governance (Xie, 2015). Exploring whether the environment is a security issue, Warner (2004) has suggested that countries do not necessarily have to respond to threats by combat, which results from 'fear'. Instead, political authorities may be 'inspired by desire' to seize such situations as 'opportunities' (Warner, 2004: 10). The author further suggests that as a result of varying mindsets in situations of extreme events, political authorities may well take chances to transform dangerous situations to profits or beneficial political assets. This seems true in the case of China. The country has shown a tendency to view water threats as opportunities to improve foreign relations and possibly political asset bargaining. As analysed in Chapters 5 and 8, water has become a politicized issue and part of a public policy issue. China's responses to disagreements on the Mekong and Amur Rivers indicate that diplomatic initiatives tackling disagreements were also taken as opportunities to promote economic cooperation with the neighbouring states.
2 On some occasions, water insecurity or water-related disasters may be viewed as a conventional security concern. It is noteworthy that China's perception of water constantly changes. Water insecurity may initially appear as a subordinate issue that is separated from conventional concern. However, in a complex environment that sees existing political tensions and unstable foreign relations, the effects of such agreements become uncertain. Despite China's physical water security level remaining relatively

> stable, its diplomatic cooperation over the GBM is largely dependent on India's responses. Apart from historical mistrust between the two countries, India is also taking a risk by assuming a more muscular approach towards China which promotes the water dispute as a core bilateral policy concern. Hence, water has come to be viewed as a bargaining chip in the pursuit of strategic security interests. In other words, the two countries' interactions have exerted more pressure for China and led to the negative transformation of China's diplomacy. As a result, the intensity of water-sharing becomes more heated in the context where there are potential defence concerns, and water may become a security concern.

Therefore, China's security agenda interplays with the changing ways in which Chinese authorities have perceived water. Consequently, the different perceptions may lead to dramatically different behaviours. China's changing perception of water is likely to be linked to the nature of China's policy processes. We can conclude that, in certain conditions, water diplomacy is likely to be vested with political interests, including industrializing states or unstable societies where policy processes lack robust democratic practices (Fischhendler, 2008). In the case of China, central government agencies represent the dominant authority to govern foreign transboundary rivers. Under its restrictive political system, such issues are most likely to be classified as politically sensitive, and the policy process is rather obscured from the public. Consequently, decision-making on water diplomacy is subject to the preferences of policymakers, whose perception is likely to constantly change. Therefore, it is no surprise that China's water diplomacy is complex and can lead to unintended, as well as intended, outcomes.

## Notes

1 ASEAN refers to a coalition of ten countries in the South-East Asia region. Originally including Indonesia, Malaysia, the Philippines, Singapore and Thailand, the organization's membership has expanded to include Brunei, Cambodia, Laos, Myanmar and Vietnam.

2 These are situations where China's water resource exploitation of river basins reaches 40 per cent.

## References

Brochmann, M. and Hensel, P. 2009. Peaceful management of international river claims, *International Negotiation*, 14(2): 393–418.

Cronin, A.A., Prakash, A., Priya, S. and Coates, S. 2014. Water in India: Situation and prospect, *Water Policy*, 16(3): 425–441.

Feng, Y., He, D. and Wang, W. 2015. Identifying China's transboundary water risks and vulnerabilities: A multidisciplinary analysis using hydrological data and legal/institutional settings, *Water International*, 40(2): 328–341.

Fischhendler, I. 2008. Ambiguity in transboundary environmental dispute resolution: The Israel–Jordanian water agreement, *Journal of Peace Research*, 45(1): 91–110.

Gerlak, A.K. and Grant, K.A. 2009. The correlates of cooperative institutions for international rivers, in T.J. Volgy, Z. Šabič, P. Roter and A.K. Gerlak, eds, *Mapping the New World Order*, New York: Wiley.

He, Y. 2008. The research on the legal problem of protecting transboundary water resources, *Journal of Natural Resources*, 23(6): 1099–1111.

Li, X., Chang, F. and Chen, J. 2011. Summary of researches on water rights and initial water rights allocation, *Journal of Yangtze River Scientific Research Institute*, 28(2): 1–9.

McIntyre, O. 2014. The protection of freshwater ecosystems revisited: Towards a common understanding of the 'ecosystems approach' to the protection of transboundary water resources, *Review of European, Comparative and International Environmental Law*, 23(1): 88–95.

Ovodenko, A. 2014. Regional water cooperation: Creating incentives for integrated management, *Journal of Conflict Resolution*, 60(6): 1071–1098.

Pu, X. 2012. Socialisation as a two-way process: Emerging powers and the diffusion of international norms, *The Chinese Journal of International Politics*, 5(4): 341–367.

Shapiro, J. 2012. *China's Environmental Challenges*, Cambridge: Polity Press.

Sørensen, G. 2010. Democracy and democratization, in K.T. Leicht and J.C. Jenkins, eds, *Handbook of Politics*, Handbooks of Sociology and Social Research Series, New York: Springer.

Springer, S. 2009. Renewed authoritarianism in Southeast Asia: Undermining democracy through neoliberal reform, *Asia Pacific Viewpoint*, 50(3): 271–276.

Wang, W. and Jia, J. 2016. Silk road economic development: Vision and path, in A. Amighini, ed., *China Dream: Still Coming True*, Novi Ligure: Edizioni Epoké.

Warner, J. 2004. *Plugging the GAP: Working With Buzan – The Illuis Dam as a Security Issue*, Occasional Paper No. 67, SOAS Water Issue Study Group, London.

Wolf, A.T., Kramer, A., Carius, A. and Dabelko, G.D. 2005. Managing water conflict and cooperation, in A.T. Wolf, ed., *State of the World: Redefining Global Security*, Washington, DC: Worldwatch Institute.

Xie, L. 2015. Political participation and environmental movements in China, in R.L. Bryant, ed., *International Handbook of Political Ecology*, Cheltenham: Edward Edgar, pp. 59–82.

Xie, L. Zhang, Y. and Panda, J. Forthcoming. Mismatched diplomacy: China–India water relations over the Ganges-Brahmaputra-Meghna River Basin, *Journal of Contemporary China*.

Zou, K. 2014. Realizing sustainability in the South China Sea, in S. Wu and K. Zou, eds, *Non-Traditional Security Issues and the South China Sea: Shaping a New Framework for Cooperation*, Farnham: Ashgate, pp. 207–224.

# 10 Policy recommendations

China's water diplomacy features versatile and changing characteristics. Along with its water diplomacy, China has, to some degree, accommodated the concerns of its neighbouring riparian states, and is likely willing to commit to developing trust and collaboration over shared water management. However, this process has not been smooth. Despite that water diplomacy has been developed as China's response to disagreements, as discussed in Chapter 2, the country has not yet formed a systematic approach to resolving disagreements with its neighbouring riparian states. In this volume, the rationale and features of China's water diplomacy have been analysed, and the outcomes of such activities are not always positive. Problems have begun to emerge in China's dealings with its neighbouring riparian states. With international water norms being problematic and insufficiently acknowledged in the global South (Salman, 2007; Uprety and Salman, 2011), China is in a tricky position where it is likely to be under scrutiny from its neighbours and international actors over its harvesting of transboundary water resources. It is predicted that China's water diplomacy will face great challenges and will have to undergo transformation if it is to realize its goals of serving the country's foreign policy and domestic water management interests.

This chapter further elaborates on why China's water diplomacy, despite that temporary successes may have been achieved, is likely to fail in the long run. The causes lie not only in China's rationale seen in its diplomatic initiatives, but also in the complex rationales of its riparian neighbours and in the features of the current international law on the governance of natural resources. This discussion is followed by an analysis of the dire consequences if China's continues on its current path of diplomatic cooperation in water sharing. The chapter concludes by providing policy recommendations for how to best secure benefits from the sharing of international rivers in a peaceful and sustainable way.

## China's water diplomacy and its limitations

China has conducted policy planning and management of international rivers, most of which are intensive activities that aim to exploit material interests from the water resources. To facilitate the realization of its national interests, China has developed diplomatic initiatives that involve varying degrees of institutional cooperation with its neighbouring riparian states. This cooperation is illustrated by the development of water treaties in the form of agreements or memoranda of understanding (MOUs). China provides technical support to those of its neighbouring riparian states that have not yet developed technological capabilities to manage the water resources of the international rivers. A clear concern in China's international water policy is peacemaking and the prevention of ecological disasters such as flooding, drought and the degradation of water quality. However, in most of China's cooperative initiatives, developing systematic water management remains a low priority.

China also supports and adheres to some of the key principles of global water norms such as the exchange of data and information with relevant states. China commits to forming water data-sharing agreements with downstream countries. In many cases, information provision and data-exchange agreements have been adopted. It is noted, however, that the scope of water information and hydrological data disclosed is under strict scrutiny from Chinese authorities. As discussed in Chapter 4, although the mechanism of information provision may satisfy certain aspects of water management for China's riparian neighbours, to what extent it facilitates communication and substantive cooperation remains in question.

### *Feasibility of China's water diplomacy*

China's current diplomatic initiatives in water management have shown an imbalance. Most of the riparian states are interested in working with China to collectively benefit from water resource management projects, such as dam construction. These states are likely to receive economic benefits from such projects, such as overcoming energy poverty and possibly other sustainable growth and development outcomes (Urban *et al.*, 2015). However, water management requires technological and financial inputs, which have become identified as obstacles for most of the riparian states. Although China is willing to act as an investor and the provider of institutions and technologies, the neighbouring states' national authorities may lack a willingness and capacity to conduct responsible and efficient environmental governance.

The issue of water cooperation represents a complicated arena where domestic politics and international relations intersect. At the international level, nation states are the official diplomatic symbol of their country but these states may fail to perform their obligations in the exploitation of water resources. For instance, national governments in the global South are often identified as prioritizing their own interests and are often simply keen to take China's investments for exploiting hydropower generation (Liebman, 2005; Zha, 2015). It is recognized that these governments are likely to fail to comply with environmental laws, often resulting in multifaceted injustices from the damming activities (Fox and Sneddon, 2007). In particular, in countries where there is weak governance, the national government may have ignored, manipulated or silenced local interests, and to be responsible for failed international cooperation efforts in the management of international river resources.

Despite China's commitment to cooperative activities, problems have arisen that result from host countries' limited capabilities. For riparian countries, improved environmental management not only results from assistance of hardware but also policy and technology transfers that support the establishment of a multilevel governance structure (Urban *et al.*, 2015). In such a context, there is a strong possibility of intra-state conflicts over the management of international rivers. Scholarship suggests that most of China's neighbouring riparian states have developed limited technologies, human resources and institution building. For instance, on the management of the Mekong River, the Lower Mekong riparian states have shown a weak capacity to tackle the fisheries and biodiversity protection issues associated with water management projects such as dam construction (Im, 2010; Urban *et al.*, 2015).

Furthermore, information disclosure on transboundary rivers is limited (see Nickum, 2008). In Asia, it is widely recognized that there is a lack of data on the reasons for the region's weak governmental capacity in water management (Grumbine and Xu, 2011). Due to limited capabilities in domestic water resource management, these countries are often unable to effectively conduct transboundary water resource management (UNEP, 2012). For collective management projects to succeed, they require sociopolitical changes at different governing levels and to show consent on water management policies (UNEP, 2012). Failing this, subnational political processes may be disturbed and this may lead to transboundary water conflicts (Lorenzoni *et al.*, 2015). In a context where riparian states are experiencing industrialization, without sufficient technology and know-how transfer their domestic politics may become unstable. Hence, much is required for China to conduct technology and know-how transfers that may involve more coordinated efforts from both sides, and the financial commitment may go well beyond the scope of the actual project. Potentially, such cooperative activities have the potential to financially drain China.

In this context, China faces high risks from such preferred benefit-sharing activities. Without developing a decentralized governance structure, river sharing may become a potential source of intra-state conflict. As shown in the withdrawal of plans for the Myitsone Dam (see Chapter 5), the political repercussions have had a serious impact on foreign relations between China and Myanmar. The Chinese authorities have underestimated the scope and scale of the work to conduct collaborative projects, which may prove not only financially infeasible, but also politically risky.

## *Voluntary deeds and ethical considerations*

In the political philosophy literature, nation states' exercise of their permanent sovereignty rights over resources is another contentious issue. While national sovereignty seems to go hand in hand with full national control over natural resources found in a state's territory, recent arguments suggest that both sovereignty-based rights claims and control over resources are conditional and relational in nature. The question thus arises whether states need to fulfil certain responsibilities and to discharge certain obligations so as to legitimately claim full control over their natural resources. This ties in with increased efforts made at global level that aim to strengthen soft international laws and to further flesh out the normative landscape surrounding concerns for transboundary issues, such as natural disaster risks, environmental degradation and dangerous anthropogenic climate change. It is in this context that a critical debate is required on the effect of sovereign control over water resources with regard to environmental sustainability.

China has shown its reservations in adopting global water norms, among which the UNWC and the UNECE are the most influential laws on the governance of transboundary water resources.[1] China disagrees with these conventions on the grounds that they impede the country's sovereign territorial rule and thus its ability to manage its water resources (He *et al.*, 2014). Concerns lie in China's joining global water norms, which raises questions on the best approach for China to exercise its sovereignty. To China, international law has not provided explicit guidance on the protection of individual countries' rightful interests. It agrees with UN environmental law that suggests nation states are rightfully entitled to appropriate their resources and the global water norms indicating that countries should not 'harm' other countries when appropriating water resources. But at the same time, the Chinese authorities have expressed doubts about how this principle is interpreted. Chinese policymakers and experts have developed their own understandings of how environmental 'harm' can be done when a country exercises its sovereign rights. China's interpretation of 'no harm' emphasizes that it implements scientific policies for the management of international rivers. At the same time, the Chinese have contested the global norms, questioning what is 'harm'? Who are the rightful/legitimate actors to articulate harm and how can these actors be best represented? And what criteria are adopted to determine if 'harm' occurs?

China has also shown moral concerns in its water policies, by voluntarily fulfilling its obligation when appropriating environmental resources to avoid inflicting harm on others. Such a moral consideration primarily aims to protect human security, including for populations beyond its national borders. However, located on the upstream of a number of major international rivers, China is viewed as possessing geographical advantages, implying that it is able to maximize its self-interest by taking unilateral actions. Its water diplomacy has shown limitations in clarifying its obligations to not harm its downstream neighbours. As such assessments are often based on ethical considerations (Barnett and Adger, 2007), there are no absolute criteria to judge if states rightfully exercise their environmental rights. With respect to China's appropriation of international rivers, whether it has violated existing international norms remains an issue for which there is no easy answer.

The effects of Chinese water diplomacy remain limited, and are probably even weakened when considering how they might contribute to resolving water disputes. Despite this, in the short term, the diplomacy may have helped China to proceed with its national water resource management interests and to realize its strategic interests. In the long run, however, it has shown clear weaknesses. In some cases, such as with India, maintaining bilateral interactions through diplomatic cooperation has proved a failure. For different reasons, both parties have encountered deadlocks. China is left with little room to negotiate on the best conditions for the country to secure its rightful interests from the shared river resources.

## *Outcomes of China's water diplomacy*

The above analysis suggests that if China continues its current diplomatic practice on the issue of water-sharing, it may well fail to achieve meaningful cooperation with its neighbouring riparian states. Such failures could have disastrous environmental and political outcomes and impact regional security.

### *Unsustainable river basins*

China's water resource management plans for its shared rivers have developed interests in promoting sustainable development in remote parts of the river basins. It is suggested that these plans are likely to succeed, given that national interests in economic profits constitute the primary incentive for developing countries to participate in cooperative activities with China (Fox and Sneddon, 2007). Indeed, such a position is echoed by some of China's riparian neighbours, particularly those that have shown incentives to conduct sustainable water management. However, apart from in a small number of cases, little concern has been shown for sustainable basin-wide water management.

In China's development of collaborative water management initiatives, water quantity remains a strong focus. This is reflected in the peaceful allocations of water within each neighbouring riparian state's territory. Although ecological considerations have been shown in China's recent national policy agenda, implementing these policies remains a challenge at both the national and international levels. Hence, there has been little collaborative development in this arena.

With the variable of the nation state's water security, we are able to understand the extent to which water resources have been utilized. The findings of Yan and Jia (forthcoming) inform us that the exploitation of the Amur River in China's North-East and the Ili River in the North-West has reached alarming levels. On these two rivers, China, Russia and Kazakhstan risk irreversible effects on river basin management and are facing potentially dire impacts from their water exploitation. In the long run, if there is no basin-scale coordination, all three of these countries will suffer, although to differing degrees, the negative impacts of unsustainable management of these river basins. In this context, China faces great challenges for how to proceed with its national interests while maintaining good foreign relations.

With China's riparian neighbours on the Mekong and Brahmaputra Rivers, there are also emerging water security issues that may exacerbate and potentially endanger foreign relations. Research suggests that the downstream riparian states are experiencing worsening situations. The water use ratio in these international rivers has been significantly increasing, to such a degree that water insecurity is beginning to occur in these river basins. The situation becomes more complex with the management of international rivers, where more substantive collective actions are required for countries to realize the ecological protection of water resources. Considering that these riparian states (i.e. India and those in the Lower Mekong Countries) have shown a weak water management capacity, their reliance on the international rivers may increase rapidly, leading to equally threatening situations that will require China's attention.

### *Regional security endangered*

With China's water diplomacy increasingly challenged, the issue of water-sharing remains a threat for its international relations in Asia. The situation is exacerbated by riparian neighbours using water-sharing as a bargaining chip in the pursuit of strategic security interests. These countries have shown a tendency to adopt the rhetoric of a downstream riparian state demanding justice and equality over transboundary water management, reflecting various water norm discourses that have been developed to empower lower-stream states in negotiations (Salman, 2010; Wegerich and Olsson, 2010). There may also be domestic elections or other national interests at play, depending on the extent to which these countries rely on the shared water resources. In this complex environment where there can be existing political tensions and unstable foreign relations, China's water diplomacy is likely to encounter ongoing difficulties.

As already shown in empirical findings, China's water diplomacy with all of its riparian neighbours face many uncertainties. If no substantive collaboration is developed on water management, China's water diplomacy is unlikely to succeed in fulfilling its purposes in the long run. One strong implication is that water-sharing remains an issue that may endanger regional security. China, being a superpower in Asia, plays a significant role in the region's security and prosperity. As shown in the interaction with some of its major international rivers, China has frequently been accused of unilaterally using water resources and harming its downstream neighbours. The increasing tensions resulting from China's actions in the South China Sea further complicate the understanding of the river system as a holistic system.[2] On the one hand, discord between China and its neighbouring riparian states is not immediately flashed with China being perceived as a threat. On the other hand, triggered by tensions over the South China Sea, the sensitive geopolitical situation in Asia can hardly sustain any more friction in foreign relations.

## Policy recommendations: a transformation of China's water diplomacy

The sharing of international rivers is a complex issue. Without exception, Asia's diplomatic water cooperation has displayed complex processes of formal cooperation (Tir and Ackerman, 2009). Scholarship in water dispute resolution has shown that there is no universal best-practice approach to water-sharing, as each country has its own unique sociopolitical conditions and have thus formed a unique identity with the international rivers. China, as with its riparian neighbours in Asia, has developed complex intentions in its cooperative behaviours with its neighbouring states. Reflected in its water diplomacy, China's water interests are dynamic and so are the outcomes of its diplomacy. Although China has displayed an increasing sense of responsibility in its sharing of transboundary waters, to best secure its interests in the sharing of international rivers, the country is faced with growing expectations to take on a leadership role to promote sustainable basin-wide management. This represents a tremendous task, considering the political instability that exists in Asia and that the role of China in the region remains contentious. Despite the challenges that China faces, there is no doubt that such deeds would benefit the livelihoods and well-being of millions of people and the generations to come in Asia.

From the above analysis, it is apparent that the rationale behind China's current diplomatic cooperation for transboundary water management has revealed limitations in serving its purposes. Maintaining friendly foreign relations will prove to be only a short-term outcome if the country continues to pay little attention to water management. In the long run, the country is faced with a serious challenge to systematically improve its diplomatic initiatives if it is to secure its national interests without exacerbating strained relations with its riparian neighbours. To maximize the benefits of developing water diplomacy, China ought to transform both the content of its water diplomacy and how this content is developed with its neighbouring riparian states. A key issue is for the country to take a lead in promoting basin-wide management.

A strong leadership is essential for Asian countries to develop collective water management projects, in a region where institution-building and material capacity for water management remains limited. With international law providing little guidance to protect China's rightful actions, its unilateral moves are likely to be interpreted as harmful. Therefore, it is recommended that China takes a leadership role in transboundary river management, which would reassure the neighbouring states of its intentions in the management of the region's transboundary rivers. Another benefit of this act lies in China maintaining its international image and continued foreign policy practices, such as its 'good-neighbourly' diplomatic initiatives (see Shi, 2013; Johnson *et al.*, 2014; Ruan, 2014); and the promotion of regional stability and security.

China has shown an increasing sense of responsibility in its sharing of transboundary waters. This is based on its intention to promote benefit sharing from sustainable river basin management. As a strong political power, China is more than capable of playing this role. China provides financial assistance and technology transfers to its partner states. As it possesses a geographical advantage over its shared international rivers, it is in a unique position where negotiation on basin-scale management can be mobilized. Hence, China ought to take the lead to promote the development of basin-wide water management mechanisms.

A strong leadership requires China to act as a driving force to promote improvements in its domestic political processes and those of its neighbouring riparian states. In China, a primary recommendation is that the government transform its diplomatic cooperation processes to orient towards building basin-wide resource management. If such processes are not promptly transformed the country risks entering into serious disputes with its neighbouring riparian states, particularly given the reducing levels of water security now facing most of these states.

China's neighbouring riparian countries are faced with great challenges to effectively establish multilevel governance arrangements (Dietz *et al.*, 2003; Raadgever *et al.*, 2008). China is well-placed to take a leading role in supporting these countries. As discussed, China's leadership role is not limited to providing financial assistance. It also involves policy transfer and support for capacity building. Strong guidance is needed for riparian states to seek ways to responsibly comply with environmental laws while simultaneously meeting their domestic interests. In this process, it would be beneficial for China to promote a certain degree of decentralization in river basin management and the incorporation of multilevel governance arrangements.

In what ways China's water management and governance experiences are imitable by its neighbouring riparian states remains unclear. Although China has thousands of years of practical experience in managing water resources, its restrictive political system raises questions about how the country could facilitate policy transfers. For instance, in China's political culture, states and political authorities are vested with trust (He *et al.*, 2013), hence they play a significant role in incorporating the public and non-state actors into the country's environmental governance. Such features may not apply to other political systems. Furthermore, China is also undergoing dramatic policy changes. In particular, the management of its international rivers is recognized as requiring a stronger legal framework.

At the international level, there are difficulties for many riparian states to commit to transboundary water resource management. As previously mentioned, the national governments of China's riparian neighbours may fail to consider this commitment. It is therefore left to China to play a leading role in basin-scale management by facilitating such initiatives. China's leading role has to be demonstrated via the building of water management institutions (WMIs), including promoting basin-scale management. International river basins run through more than one nation, and so the development of basin-scale governance has to respect the natural scale of water basins rather than just the jurisdictions where the river is located.

China's emphasis on developing state-led cooperation has also shown limitations. As discussed above, weaknesses abound in state-led cooperation, particularly in the context of the global South. The Chinese authorities have been monopolizing this process, showing little interest in developing formal initiatives to incorporate non-state actors and experts. This approach undermines the purpose of trust-building. It also does not comply with the country's domestic policies on information disclosure which promote transparency in environmental governance as a recognized approach to realize sustainable development (Cai and Geng, 2006). In a WMI led by China, we recommend the incorporation of non-state actors including experts. This would increase transparency and may prove effective for the country to promote its intentions for the management of international rivers. As adopted in a few cases, incorporating national or international NGOs has proven to be a success in areas where close cooperation has been built. It has also had the effect of improving policy processes by incorporating contributions from non-state actors and scientific authorities.

When negotiating for a cooperative mechanism with its neighbouring riparian states, China should reconsider the effects of information exchanges to promote trust-building in river sharing. Information-sharing is a key feature of transboundary water governance, which incorporates information disclosure on various levels of water management as well as decentralized authority. The experiences mostly come from Western Europe or North America where democratic practices have proved to be necessary political and social conditions for cooperative mechanisms to function (Mason, 2008).

Additionally, information-sharing has been adopted as a diplomatic initiative in China's interactions with its neighbouring riparian states. With the intention of soothing the mistrust that has arisen from its appropriating of international rivers, China has included information-sharing in most of its diplomatic cooperation. However, China has mostly disclosed only partial information, limiting the success of such cooperative efforts. As discussed in Chapter 6, information-sharing has effectively complicated negotiations between China and India, hence it did not prove to be a great success.

## Notes

1 These are the UN Convention on the Law of Non-Navigational Use of International Watercourses (1997); the UNECE Convention on the Protection and Use of Transboundary Watercourses and International Lakes (1992).

2 According to the national border demarcated by the Indian government, the river basin has a minor branch that runs through South Tibet in China.

## References

Barnett, J. and Adger, W.N. 2007. Climate change, human security and violent conflict, *Political Geography*, 26(6): 639–655.

Cai, J. and Geng, Q. 2006. Study on concept and data collection system of scientific data sharing of water resources, *Journal of China Institute of Water Resources and Hydropower Research*, 4(1): 31–35.

Dietz, T., Ostrom, E. and Stern, P.C. 2003. The struggle to govern the commons, *Science*, 302(5652): 1907–1912.

Fox, C. and Sneddon, C. 2007. Transboundary river basin agreements in the Mekong and Zambezi basins: Enhancing environmental security or securitising the environment, *International Environmental Agreements: Politics, Law and Economics*, 7(3): 237–261.

Grumbine, R.E. and Xu, J. 2011. Mekong hydropower development, *Science*, 332(6026): 178–179.

He, D., Liu, C., Feng, Y., Hu, J., Ji, X. and Li, Y. 2014. Progress and perspective of international river researches in China, *Acta Geographica Sinica*, 69(9): 1284–1294.

He, G., Mol, A., Zhang, L. and Lu, Y. 2013. Public participation and trust in nuclear power development in China, *Renewable and Sustainable Energy Reviews*, 23(7): 1–11.

Im, S.S. 2010. *Research, Knowledge Transfer and Capacity Building Programs Related to Hydropower Development in Cambodia*, SPLASH-ERA-NET. Available at: http://splash-era.net/downloads/mekong_report_part2.pdf (accessed 1 May 2015).

Johnson, C., Bower, E., Cha, V., Green, M. and Goodman, M. 2014. *Decoding China's Emerging 'Great Power' Strategy in Asia*, Washington, DC: Center for Strategic and International Studies.

Liebman, A. 2005. Trickle-down hegemony? China's 'peaceful rise' and dam building on the Mekong, *Contemporary Southeast Asia: A Journal of International and Strategic Affairs*, 27(2): 281–304.

Lorenzoni, I., Benson, D. and Cook, H. 2015. Regional rescaling in UK adaptation governance: From agency to collaborative control in flood management in England? In J. Knieling, ed., *Climate Adaptation Governance Theory in Cities and Regions: Theoretical Fundamentals and Practical Evidence*, Chichester: Wiley.

Mason, M. 2008. The governance of transnational environmental harm: Addressing new modes of accountability/responsibility, *Global Environmental Politics*, 8(3): 8–24.

Nickum, J.E. 2008. The upstream superpower: China's international rivers, in O. Varis, C. Tortajada and A.K. Biswas, eds, *Management of Transboundary Rivers and Lakes*, Berlin: Springer, pp. 227–244.

Raadgever, G.T., Mostert, E., Kranz, N., Interwies, E. and Timmerman, J.G. 2008. Assessing management regimes in transboundary river basins: Do they support adaptive management? *Ecology and Society*, 13(1): art. 14. Available at: www.ecologyandsociety.org/vol13/iss1/art14/ (accessed 10 July 2017).

Ruan, Z. 2014. Zhongguo xuyao goujian zengyang de zhoubian, *Guoji Wenti Yanjiu*, 2: 11–26.

Salman, S.M.A. 2010. Downstream riparians can also harm upstream riparians: The concept of foreclosure future uses, *Water International*, 35(4): 350–364.

Shi, Y. 2013. Zhongguo zhoubian xingwei zhong cengyoude 'shengli zhuyi': Dongneng he juece fuzaxing, *Xiandai Guoji Guanxi*, 10: 3–5.

Tir, J. and Ackerman, J.T. 2009. Politics of formalized river cooperation, *Journal of Peace Research*, 46(5): 623–640.

UNEP (United Nations Environment Programme). 2012. *Status Report on the Application of Integrated Approaches to Water Resources Management*, Nairobi: UNEP. Available at: www.un.org/waterforlifedecade/pdf/un_water_status_report_2012.pdf (accessed 17 August 2017).

Uprety, K. and Salman, S.M.A. 2011. Legal aspects of sharing and management of transboundary water in South Asia: Preventing conflicts and promoting cooperation, *Hydrological Sciences Journal*, 56(4): 641–661.

Urban, F., Siciliano, G., Sour, K., Lonn, P.D., Tan-Mullins, M. and Mang, G. 2015. South–South technology transfer of low-carbon innovation: Large Chinese hydropower dams in Cambodia, *Sustainable Development*, 23(4): 232–244.

Wegerich, K. and Olsson, O. 2010. Late developers and the inequity of 'equitable utilization' and the harm of 'do not harm', *Water International*, 35(6): 707–717.

Yan, J. Forthcoming. Assessment of China's transboundary water resources and water use, doctoral thesis, Chinese Academy of Sciences, Beijing.

Zha, D. 2015. A political ecology of hydropower development in China, in W. Jisi, ed., *Review of China's International Strategies*, Beijing: Center for International and Strategic Studies, Peking University.

# Index

Page numbers in *italics* denote tables, those in **bold** denote figures.